3rd International Conference and Young Scientist School on Magnetic Resonance Imaging in Biomedical Research 2016

Journal of Physics: Conference Series Volume 886

Novosibirsk, Russia
23 – 26 November 2016

ISBN: 978-1-5108-4789-7
ISSN: 1742-6588

Printed from e-media with permission by:

Curran Associates, Inc.
57 Morehouse Lane
Red Hook, NY 12571

Some format issues inherent in the e-media version may also appear in this print version.

Printed by Curran Associates, Inc. (2017)

For permission requests, please contact the Institute of Physics
at the address below.

Institute of Physics
Dirac House, Temple Back
Bristol BS1 6BE UK

Phone: 44 1 17 929 7481
Fax: 44 1 17 920 0979

techtracking@iop.org

Additional copies of this publication are available from:

Curran Associates, Inc.
57 Morehouse Lane
Red Hook, NY 12571 USA
Phone: 845-758-0400
Fax: 845-758-2633
Email: curran@proceedings.com
Web: www.proceedings.com

Table of contents

Volume 886

3rd International Conference and Young Scientist School "Magnetic resonance imaging in biomedical research"

23–26 November 2016, Novosibirsk, Russian Federation

Accepted papers received: 14 August 2017
Published online: 29 August 2017

Preface

Papers

The Third International Conference and Young Scientist School "Magnetic Resonance Imaging in Biomedical Research"

A.V. Naumova [1,2], M.Yu. Khodanovich[1,2], V.L.Yarnykh[1,2]

[1] Department of Radiology, University of Washington, Seattle, WA, USA
[2] Research Institute of Biology and Biophysics, Tomsk State University, Tomsk, Russia

ABSTRACT

The Third International Conference and Young Scientist School "Magnetic Resonance Imaging in Biomedical Research" was held in Novosibirsk, Russia, on November 24-26, 2016. The Conference was held within the framework of the International Congress on Clinical and Translational Neuroimaging organized by the International Tomography Center of the SB RAS, Institute of Medicine and Psychology at Novosibirsk State University (NSU), Institute of Cytology and Genetics of the SB RAS, and Tomsk State University (TSU).

The Conference focused on magnetic resonance imaging (MRI) applications for biomedical research. The main goal was to bring together major scientists, clinical researchers and developers of new MRI techniques to bridge the gap between clinical/research needs and advanced technological solutions. The Conference encouraged research and development in basic and clinical MR science and its application in health care. It also had an educational purpose to promote understanding of cutting-edge MR developments. The Conference provided an opportunity for researchers and clinicians to present their recent theoretical developments, practical applications, and to discuss acute problems.

The program of the Conference was divided into three main topics. The first day of the Conference was devoted to educational lectures on the fundamentals of MRI physics and image acquisition/reconstruction techniques including recent developments in quantitative MRI. The second day focused on developments and applications of new contrast agents. Multinuclear and spectroscopic acquisitions as well as functional MRI were presented on the third day of the Conference.

We would like to highlight the main developments presented at the Conference and introduce the prominent speakers. The keynote speaker of the conference Dr. Vasily Yarnykh (University of Washington, Seattle, USA) demonstrated a method of macromolecular proton fraction and its application in clinical and preclinical studies. Professor Yury Pirogov (Lomonosov Moscow State University) presented development of multinuclear magnetic resonance imaging in biomedical experiments. Dr. Marina Khodanovich (Tomsk State University) described histological validation of macromolecular proton fraction mapping on the model of cuprizone-induced demyelination.

We express our thanks to the scientific committee, the local organizing committee and Tomsk State University for giving the opportunity to share scientific ideas and up-to-date developments at the Conference, and the Russian Science Foundation (project № 14-45-00040) and Biovitrum Ltd. for financial support.

IOP Conf. Series: Journal of Physics: Conf. Series **886** (2017) 011002 doi:10.1088/1742-6596/886/1/011002

Peer review statement

All papers published in this volume of *Journal of Physics: Conference Series* have been peer reviewed through processes administered by the proceedings Editors. Reviews were conducted by expert referees to the professional and scientific standards expected of a proceedings journal published by IOP Publishing.

IOP Conf. Series: Journal of Physics: Conf. Series **886** (2017) 012001 doi:10.1088/1742-6596/886/1/012001

Calculation of optimal parameters for ^{19}F MRI

N Anisimov[1], M Gulaev[1], O Pavlova[1], D Fomina[2], V Glukhova[2], S Batova[2] and Yu Pirogov[2]

[1]Faculty of Fundamental Medicine, Lomonosov Moscow State University, Moscow Russia

[2]Faculty of Physics, Lomonosov Moscow State University, Moscow, Russia

E-mail: anisimovnv@mail.ru

Abstract. This paper presents a method for optimizing the parameters of the scanning pulse sequences for MRI in relation to objects with a wide NMR spectrum. In this case, a broadband excitation of the spin system is difficult because of hardware limitations. It is proposed to apply the selective excitation, the optimum parameters of which are calculated by an algorithm that uses information concerning the NMR spectrum. The method is especially useful for ^{19}F MRI of fluorocarbons.

1. Introduction. Purpose

There is a problem of obtaining MR images of object with a wide NMR spectrum. This is especially true for the ^{19}F MRI [1,2]. The wider is the spectrum the higher are requirements for transmitter power, gradient system strength, data sampling rate and receiver sensitivity. These requirements grow with increasing magnetic field. To solve this problem one needs to use narrow-band (frequency selective) excitation of the spin system. In this case it is necessary to optimize the scanning parameters - choose the excitation frequency and spectrum width. We made software that solves this task. It uses information about a specific NMR spectrum and calculates the scanning parameters which give the maximum MR signal. We used calculation results to optimize pulse sequences parameters in our *in vivo* MRI studies which were carried out at 0.5- and 7-T scanners [3,4].

2. Materials and methods

The idea of the method is to calculate the signal of nuclear induction S(t) for different variants of frequency selective excitation of the spin system - excitation frequency f_0, spectral width of excitation pulse Δ - and to select the optimal one. Search for optimal parameters (t, f_0, Δ) is made within the intervals defined by technical limitations. It is assumed that they are known for the equipment on which it is planned to carry out MRI. Sorting out of variants is realized by specialized software that makes calculations in batch mode and gives the excitation parameters in which the maximum MR signal is realized.

Scheme of the method is shown on Figure 1. High resolution NMR spectrum F(f) is multiplied by a Gaussian function $G(f)=G(f,f_0,\Delta)=\exp(-(2\times(f-f_0)/\Delta)^2)$. The result undergoes to inverse Fourier transform (IFT) that gives value $S(t)=S(t,f_0,\Delta)$. Then its magnitude $|S(t)|$ is calculated.

It is assumed that $|S(t)|$ determines the brightness of the pixels on the MR image. Therefore, the values of "t" which give extremes of the $|S(t)|$ serve as a reference points for setting the optimal TE value of gradient echo pulse sequence (GE). The value of |S(t)| at t ~ 0 determines the MR signal for spin echo pulse sequence (SE).

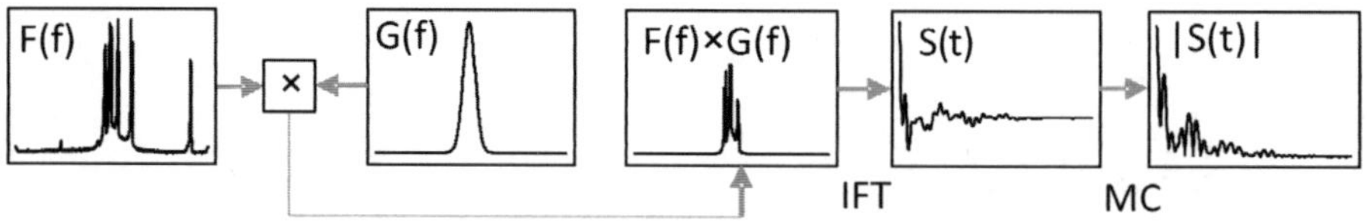

Figure1. The scheme of calculating the MR signal by the selective excitation.

The program creates a 3D array of |S| = S(t,f_0,Δ), and then searches for the two sets of parameters (t,f_0,Δ). One of them gives the maximum value for |S|, the second - for $|S|\times\Delta^{-1/2}$. The latter value is proportional to the signal-to-noise ratio (|S|/N) as the root mean square amplitude of the thermal noise N is proportional to $\Delta^{1/2}$.The range of variable parameters of frequency selective excitation is defined considering hardware limitations. The gradient system strength determines the minimum values of TE, the transmitter power and the permissible radiofrequency load determine maximum value of Δ, as it is inversely proportional to the pulse length.

The method is most suitable for pulse sequences, which use Gaussian pulse (its length τ cut-off of 10% is calculated by the formula: $\tau = 2.74/\Delta$) with small flip angle (FA) [5]. These requirements can be easily fulfilled in the GE method. But the results of calculations can be demanded for the SE technique where the flip angles are 90° and 180°. However, in this case, the results of calculation should be used only for a rough estimate of the optimum parameters. In general, for accurate calculations instead of the Gaussian function G(f) excitation profile calculated from the Bloch equations should be used, [3]. In addition, it is necessary to take into account differences in the longitudinal relaxation times for different spectral lines. Besides, to clarify the optimal TE for the SE method effects of transverse relaxation and J-modulation should be taken into consideration, therefore the experimental dependence of |S| on TE should be measured.

3. Results

The method has been tested for fluorocarbon drug Perftoranum® [6]. Figure 2 shows its ^{19}F NMR spectra which were obtained in the fields of 0.5 and 7 Tesla using the Bruker's scanners – Tomikon S50 and BioSpec 70/30 USR. In addition to spectrum data we introduced into our program the limit values for Δ and TE, to reduce the range of variable parameters.

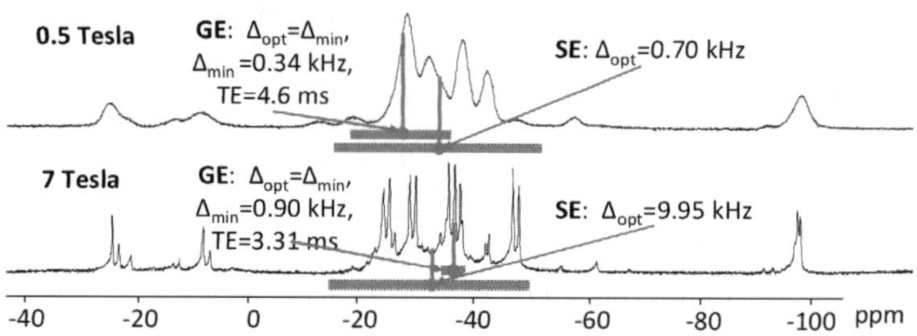

Figure 2. Spectra of Perftoranum® at fields 0.5 and 7 T and the optimal parameters for the selective excitation of the spin system, defined according to the calculating algorithm.

Figure 2 shows the results of calculation - parameters TE, f_0 and Δ, giving the maximum value of |S|/N for the GE and the SE techniques. Vertical red lines indicate the points of the excitation frequencies with respect to the NMR spectrum. The lengths of the horizontal lines correspond to the optimum values of Δ. As for the GE method, the value of Δ should be as small as possible. This value

IOP Conf. Series: Journal of Physics: Conf. Series **886** (2017) 012001 doi:10.1088/1742-6596/886/1/012001

is determined by technical limitations. For the method of SE optimum values Δ correspond roughly to the width of the spectrum part in which the most intense lines are concentrated.

Calculated optimal parameters are of practical interest if a small deviation of them from the optimum values does not lead to a significant decrease of the MR signal. To estimate the stability of the optimal mode of selective excitation, we analyzed the behavior of functions |S| and |S|/N in the vicinity of their maximum values when two of the three parameters are fixed. Figure 3 shows the corresponding calculated (blue) and experimental (red) graphs for the GE and the SE modes.

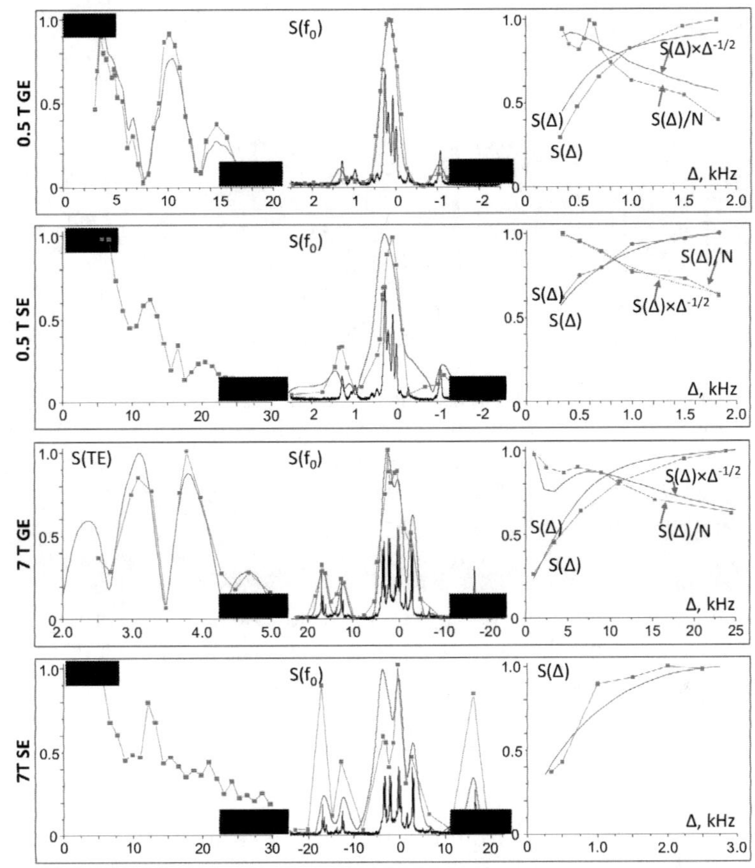

Figure 3. Calculated (blue) and experimental (red) curves of |S| and |S|/N for GE and SE modes. Measurements were carried out on 0.5- and 7-T MRI scanners.

For the SE mode, the calculated values of S(t) are meaningful only for t ~ 0. Therefore experimental graphs |S(TE)| for the SE mode are shown to reveal J-modulation and T_2 decay. These factors should also be taken into account when calculating the optimum values of TE. Common values are: TR = 0.5 s, and FA = 30^0 for the GE mode. Attention was drawn to the inadmissibility of the chemical shift artifacts on the MR image. Therefore, for the study |S| = |S(Δ)|/N we installed sampling rate - receiver bandwidth (BW) - equal to $\Delta \times N_f$, where N_f is number of frequency encoding steps. Typically, N_f was 64. In all other experiments we installed BW at maximal possible value - up to 0.2 and 1 MHz for 0.5 and 7 T scanners, accordingly.

It can be seen that the experimental graphs, in general, coincide closely with the calculated values. The largest discrepancy is noted only for the graph $S(f_0)$ for the SE mode. This is due to the fact that TR = 0.5 s was comparable with the longitudinal relaxation times T_1, and these times are significantly various for the peaks belonging to the different components of the fluorocarbon compound. For the method GE this factor is not significant, because FA is small. It may be noted that curves $|S(\Delta)|$ increase monotonically with increasing Δ. But for dependence of the signal-to-noise ratio ($|S(\Delta)|/N$) the situation is reversed. This is due to the fact that the NMR spectrum is not a band but is a set of narrow lines. Therefore with the increase of Δ, the total induction signal is growing slower than the thermal noise whose spectrum is continuous and its amplitude increases $\sim \Delta^{1/2}$. This does not exclude the presence of local extrema near Δ, comparable to the spectral width of the zone in which the most intense peaks are concentrated.

We used the results of the calculations when setting the parameters of MRI scanning *in vivo* studies using laboratory animals (7 T) and a human body (0.5 T) – figure 4. The combination of ^{1}H and ^{19}F images help to clarify the localization of fluorine relatively to anatomical structures. Sometimes the presence of fluorine can be detected on the ^{1}H MR image by loss of signal (the rightmost fragment).

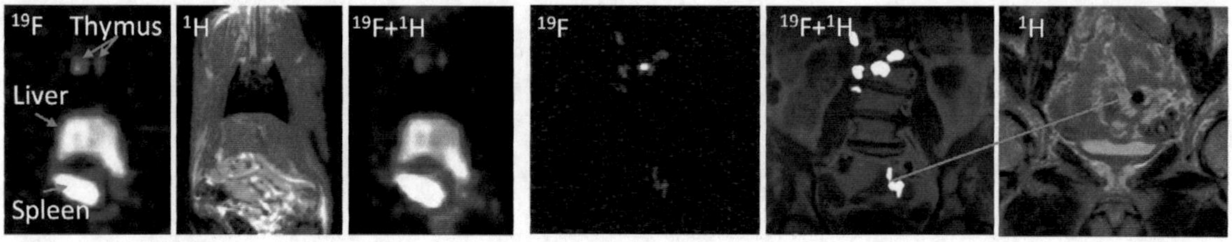

Figure 4. Left three fragments: ^{19}F MRI (SE) shows accumulation of the drug Perftoranum® in the liver, spleen and thymus after intraperitoneal injection of the drug into the body of a rat. Right three fragments show the moving capsule filled by perfluorodecalin (main component of Perftoranum®) in the human gastrointestinal tract. ^{19}F MRI (GE) is the sum of the images obtained at intervals of 0.5 to 1 hour. The total time of the study - 8 hours.

4. Discussion. Conclusion

The proposed method allows in relatively simple manner to calculate the optimal parameters of MRI scanning of objects with complex NMR spectrum considering hardware resources. The method can be adapted for more accurate calculations by using excitation profile calculated from the Bloch equations instead of the Gaussian function. Method efficiency is improved by adding information about the parameters of relaxation and spin-spin coupling constants.

Acknowledgements
This work was supported by the grant from the Ministry of Education and Science of the Russian Federation No. 14.604.21.0060 (RFMEFI60414X0060). The authors acknowledge the support of Russian Science Foundation grant No 14-45-00040-П for manuscript preparation and publication.

References

[1] Tirotta I, Dichiarante V, Pigliacelli C, Cavallo G, Terraneo G, Bombelli F B, Metrangolo P and Resnati G 2015 *Chem. Rev.* **115** (2) 1106
[2] Jesu's Ruiz-Cabelloa, Barnett B P, Bottomley P A and Bulte J W M 2011 *NMR in Biomed.* **24** 114
[3] Gulyaev M V, Gerviz L L, Ustynyuk Y A, Anisimov N V, Pirogov Yu A and Khokhlov A R 2013 *Journal of radioelectronics* No 8 - http://jre.cplire.ru/jre/aug13/8/text.html *(in Russian)*
[4] Volkov D V, Pavlova O S, Gulyaev M V, Anisimov N V and Pirogov Yu A 2014 *Journal of radioelectronics* No 11- http://jre.cplire.ru/jre/aug14/11/text.html *(in Russian)*
[5] Bruker-BioSpin 2003 *Shaped Pulses. PV 3.0.1 Online Manual* D-5-8
[6] Maevsky E I and Gervits L L 2008 *Supplement of Chimica Oggi/CHEMISTRY TODAY, Focus on Fluorine Chemistry* **26** (3) 8

Corrigendum: Calculation of optimal parameters for ^{19}F MRI

J. Phys. Conf. Ser **886** (2017) 012001

N Anisimov[1], M Gulaev[1], O Pavlova[1], D Fomina[2], V Glukhova[2], S Batova[2] and Yu Pirogov[2]
[1]Faculty of Fundamental Medicine, Lomonosov Moscow State University, Moscow Russia
[2] Faculty of Physics, Lomonosov Moscow State University, Moscow, Russia.
E-mail: anisimovnv@mail.ru

Page 3, Figure 3:

The published version of figure 3 is corrupted. This corrigendum contains corrected version of the image.

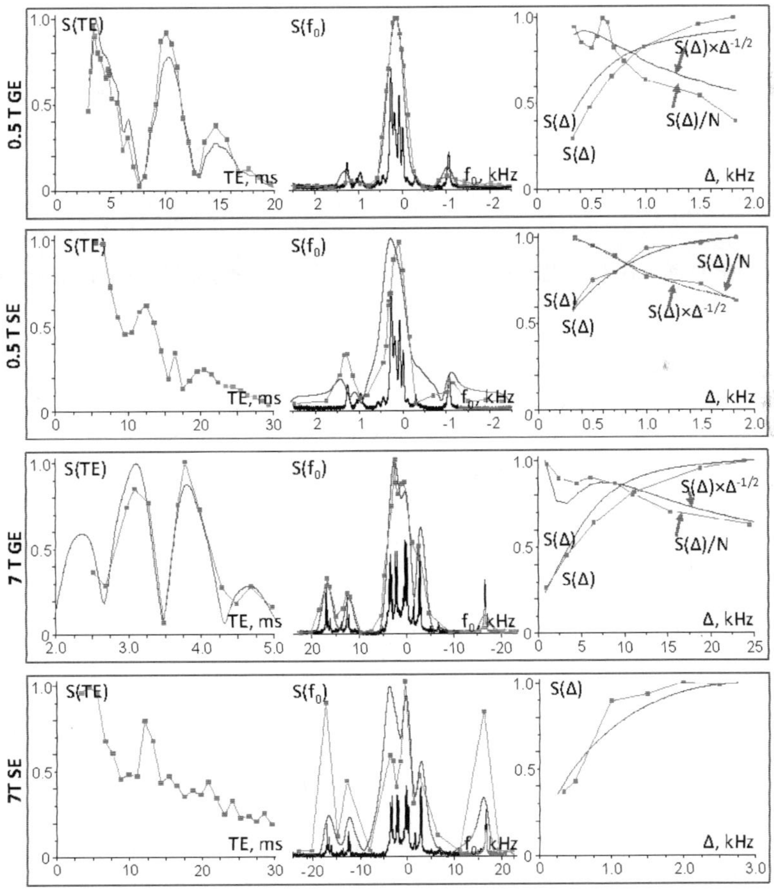

Figure 3. Calculated (blue) and experimental (red) curves of |S| and |S|/N for GE and SE modes. Measurements were carried out on 0.5 and 7 T MRI scanners.

IOP Conf. Series: Journal of Physics: Conf. Series **886** (2017) 012002 doi:10.1088/1742-6596/886/1/012002

Role of eNOS in water exchange index maintenance-MRI studies

D Atochin[1,2,3], M Litvak[1,2], S Huang[4], Y R Kim[4], P Huang[1]
[1] Cardiovascular Research Center and Cardiology Division, Massachusetts General Hospital, Harvard Medical School, Charlestown, Massachusetts, USA
[2] RASA Center in Tomsk, Tomsk Polytechnic University, Tomsk, Russia
[3] Research Institute of Biology and Biophysics, Tomsk State University, Tomsk, Russia
[4] Athinoula A. Martinos Center for Biomedical Imaging/Radiology, Massachusetts General Hospital, Harvard Medical School, Charlestown, Massachusetts, USA

E-mail: atochin@cvrc.mgh.harvard.edu

Abstract. Stroke studies employ experimental models of cerebral ischemic and reperfusion injury in rodents. MRI provides valuable supravital data of cerebral blood flow and brain tissue damage. This paper presents MRI applications for cerebral blood flow research in mice lines with impaired nitric oxide production by endothelial nitric oxide synthase. Our data demonstrates that specific modifications of MRI methodology in transgenic mouse models help to evaluate the role of eNOS in the brain-blood barrier function.

1. Introduction

According to the World Health Organization, 15 million people suffer from stroke worldwide each year. Of these, 5 million die and another 5 million are permanently disabled. Stroke remains the third leading cause of death in the United States, Europe and Russia. More than 140,000 people die each year from stroke in the United States; 650,000 and 265,000 – in Europe and in Russia respectively [1].

Experimental stroke studies employ methods of modeling brain ischemia and reperfusion injury in rodents with cerebrovascular surgical approaches. Knockout and knockin mouse lines allow targeting of specific enzymes to study its roles in development of stroke injury [2, 6]. Nitric oxide (NO) derived by endothelial NO synthase (eNOS) is important for regulating vascular tone, maintaining blood pressure, and protection against stroke [2-6]. eNOS was proven to be a key signal-transduction enzyme responsible for endothelium-dependent vasodilation, and proper cerebral blood flow [7]. Mice lacking eNOS (eNOS KO) demonstrated significant hypertension and larger brain infarction upon cerebral ischemia [8]. eNOS is acivated in response to shear stress, insulin, insulin-like growth factor, estrogen, statins [9-12], PPAR-gamma, leptin [13, 14] and adiponectin [15]. Phosphorylation of eNOS at serine 1176 (S1176) results in increased production of NO [16, 17] and modulates vascular reactivity [18]. Substitution of S1176 by aspartate (S1176D) mimics the enhanced catalytic activity caused by phosphorylation [19, 20] and increased NO production [4, 9, 10].

2. Materials and methods

We generated knock-in mice carrying phosphomimetic (S1176D) mutation in the eNOS gene [21]. To generate S1176D knockin mice the eNOS gene was replaced with a mutant gene carrying the S1176D mutation [10]. S1176D mice were back-crossed on the C57BL6 background. eNOS protein levels in aortic tissues were similar in WT and S1176D mice. We demonstrated that S1176D mice produced more NO as compared with WT mice [21].

In this paper, we aimed to demonstrate differences in MRI-derived water exchange index (WEI), a novel indicator of vascular integrity via measuring the rate of water movement across the BBB, between wild type (WT) mice and mice with impaired eNOS function.

Eight WT (C57BL/6), five eNOS KO mice and three S1176D knockin mice were used in this study. A 3D SPGR (spoiled gradient echo) pulse sequence was used for quantifying transvascular water exchange using a 9.4T scanner (Bruker Biospin). Before and after intravenous administration of a strictly intravascular contrast agent Gd-PGC (MW~500,000Da), 3D volume images (Matrix: 64 x 64 x 128, FOV: 1.5 x 1.5 x 2.5 cm) were collected with TR/TE = 30/3 ms with varying flip angle = 30 and 90° degrees. Additional 3D-SPGR scans with TR/TE = 30/5, 7 ms at a fixed flip angle of 30° were also performed to calculate changes in T2* due to Gd-PGC. Water exchange index (WEI) was defined as the following: WEI = Vapp 30 /Vapp 90 [22, 23].

3. Results

Baseline WEI significantly increased in the eNOS KO mice (2.08±0.53) as compared with WT (1.43±0.20, Mean±SD, P<0.05), indicating higher BBB water permeability. Furthermore, baseline WEI significantly decreased in the S1176D mice (1.29±0.20) as compared with eNOS KO mice. These results suggest that the impaired eNOS affects vascular function (i.e., water permeability) even under a normal physiological condition. Understanding such abnormalities in baseline cerebrovascular parameters in eNOS mutant mouse models will further aid the investigation of eNOS functions under pathological conditions.

Acknowledgments

The part of reported study devoted to experimental surgery in mice was supported by the Russian Science Foundation, Grant No. 17-15-01111

References

[1] World Health Organization 2002 World Health Report 2002: Reducing Risks, Promoting Healthy Life *Geneva: World Health Organization* 9–10

[2] Atochin D N, Huang P L 2011 Role of endothelial nitric oxide in cerebrovascular regulation Curr. Pharm. Biotechnol. 12 (**9**) 1334–42

[3] Fok H, Jiang B, Clapp B, Chowienczyk P 2012 Regulation of vascular tone and pulse wave velocity in human muscular conduit arteries: selective effects of nitric oxide donors to dilate muscular arteries relative to resistance vessels *Hypertension* **60** (5) 1220-5

[4] Schleicher M, Yu J, Murata T, Derakhshan B, Atochin D N, Qian L, Kashiwagi S, Di Lorenzo A, Harrison K D, Huang P L, Sessa W C 2009 The Akt1-eNOS axis illustrates the specificity of kinase-substrate relationships in vivo *Sci. Signal.* **2** (82) 23-24

[5] Atochin D N, Schepetkin I A, Khlebnikov A I, Seledtsov V I, Swanson H, Quinn M T, Huang P L 2016 A novel dual NO-donating oxime and c-Jun N-terminal kinase inhibitor protects against cerebral ischemia-reperfusion injury in mice *Neurosci Lett.* **618** 45-9

[6] Atochin D N, Huang P L 2010 Endothelial nitric oxide synthase transgenic models of endothelial dysfunction *Pflugers Arch.* **460** (6) 965-74

[7] Gulati P, Singh N, Muthuraman A 2014 Pharmacologic evidence for role of endothelial nitric oxide synthase in neuroprotective mechanism of ischemic postconditioning in mice *J. Surg. Res.* **188(1)** 349–60

[8] Huang P L, Huang Z, Mashimo H, Bloch K D, Moskowitz M A, Bevan J A, Fishan M C 1995 Hypertension in mice lacking the gene for endothelial nitric oxide synthase *Nature* **377**(6546) 239-242

[9] Dimmeler S, Fleming I, Fisslthaler B, Hermann C, Busse R., Zeiher A M 1999 Activation of nitric oxide synthase in endothelial cells by Akt-dependent phosphorylation *Nature* **399** (6736) 601–5

[10] Fulton D, Gratton J P, McCabe, T J, Fontana J, Fujio Y, Walsh K, Franke T F, Papapetropoulos

A, Sessa W C 1999 Regulation of endothelium-derived nitric oxide production by the protein kinase Akt *Nature* **399** (6736) 597–601

[11] Kureishi Y, Luo Z, Shiojima I, Bialik A, Fulton D, Lefer D J, Sessa W C, Walsh K 2000 The HMG-CoA reductase inhibitor simvastatin activates the protein kinase Akt and promotes angiogenesis in normocholesterolemic animals. *Nat. Med.* **6** (9) 1004–10

[12] Snyder S H, Jaffrey S R 1999 Vessels vivified by Akt acting on NO synthase *Nat. Cell Biol.* **1 (4)** E95–96

[13] Cho D H, Choi Y J, Jo S A, Jo I 2004 Nitric oxide production and regulation of endothelial nitric-oxide synthase phosphorylation by prolonged treatment with troglitazone: evidence for involvement of peroxisome proliferator-activated receptor (PPAR) gamma-dependent and PPARgamma-independent signaling pathways *J. Biol. Chem.* **279** (4) 2499–2506

[14] Goetze S, Bungenstock A, Czupalla C, Eilers F, Stawowy P, Kintscher U, Spencer-Hansch C, Graf K, Nurnberg B, Law R E, Fleck E, Grafe M 2002 Leptin induces endothelial cell migration through Akt, which is inhibited by PPARgamma-ligands *Hypertension* **40** (5) 748–54

[15] Chen H, Montagnani M, Funahashi T, Shimomura I, Quon M J 2003 Adiponectin stimulates production of nitric oxide in vascular endothelial cells *J. Biol. Chem.* **278** (45) 45021–26

[16] Michell B J, Griffiths J E, Mitchelhill K I, Rodriguez-Crespo I, Tiganis T, Bozinovski S, de Montellano P R, Kemp B E, Pearson R B 1999 The Akt kinase signals directly to endothelial nitric oxide synthase *Curr. Biol.* **9** (15) 845–48

[17] Sorenson J, Santhanam A V, Smith L A, Akiyama M, Sessa W C, Katusic Z S 2005 Expression and function of recombinant S1179D endothelial NO synthase in human pial arteries. *Stroke* **36** (1) 158–160.

[18] Atochin D N, Wang A, Liu V W, Critchlow J D, Dantas A P, Looft-Wilson R, Murata T, Salomone S, Shin H K, Ayata C, Moskowitz M A, Michel T, Sessa W C, Huang P L 2007 The phosphorylation state of eNOS modulates vascular reactivity and outcome of cerebral ischemia in vivo *J. Clin. Invest.* **117** (7) 1961–67.

[19] McCabe T J, Fulton D, Roman L J, Sessa W C 2000 Enhanced electron flux and reduced calmodulin dissociation may explain “calcium-independent” eNOS activation by phosphorylation *J. Biol. Chem.* **275** (9) 6123–28

[20] Scotland R S, Morales-Ruiz M, Chen Y, Yu J, Rudic R D, Fulton D, Gratton J P, Sessa W C 2002 Functional reconstitution of endothelial nitric oxide synthase reveals the importance of serine 1179 in endothelium-dependent vasomotion *Circ. Res.* **90** (8) 904-910

[21] Kashiwagi S, Atochin D N, Li Q, Schleicher M, Pong T, Sessa W C, Huang P L 2013 eNOS phosphorylation on serine 1176 affects insulin sensitivity and adiposity *Biochem. Biophys. Res. Commun.* **431** (2) 284–290

[22] Kim Y R, Rebro K J, Schmainda K M 2002 Water exchange and inflow affect the accuracy of T1-GRE blood volume measurements: Implications for the evaluation of tumor angiogenesis *Magn. Reson. Med.* **47** (6) 1110–20

[23] Kim Y R , Tejima E, Huang S, Atochin D N, Dai G, Lo E H, Huang P L, Bogdanov A Jr, Rosen B R 2008 In vivo quantification of transvascular water exchange during the acute phase of permanent stroke *Magn. Reson. Med.* **60** (4) 813–21

Prospects of paramagnetic lanthanide complexes for magnetic resonance imaging, local thermosensing and diagnosing

S Babailov[1], A Akulov[2,3], M Moshkin[3] and I Koptyug[4]

[1]Nikolayev Institute of Inorganic Chemistry of the Russian Academy of Sciences, Novosibirsk, Russia
[2]Institute of Cytology and Genetics of the Russian Academy of Sciences, Novosibirsk, Russia
[3]Research Institute of Biology and Biophysics, Tomsk State University, Tomsk, Russia
[4]International Tomography Centre of the Russian Academy of Sciences, Novosibirsk, Russia

E-mail: babajlov@niic.nsc.ru

Abstract. The features of the paramagnetic properties of lanthanide (Ln) containing complexes with crown ethers, pivaloyltrifluoracetonato, and EDTA have been studied by ^{1}H and ^{19}F NMR techniques. These complexes are of interest due to the prospect of using some of them as sensors for biology and medicine. The temperature dependencies of the lanthanide-induced shifts revealed that Ln complexes might have practical importance for ^{1}H and ^{19}F NMR temperature control in solutions. It was shown *in vitro* (using NMR spectrometers and MRI scanners) that these complexes can serve as special NMR thermo-sensitive contrast reagents.

1. Introduction

MRI is routinely used in the clinic to visualize the structure and function of organs. For tumor and cardiovascular imaging in clinic and *in vivo* studies, an enhancement in contrast is an absolute requirement to improve the sensitivity and the diagnostic contents of the images. For this, several contrast agents (CAs) have been developed earlier, most of which incorporates paramagnetic gadolinium ions (Gd^{3+}). Currently, complexes of the different lanthanides (Ln, other than Gd), remain largely unexplored as probes for nuclear magnetic resonance (NMR) spectroscopy and magnetic resonance imaging (MRI).

Recent studies have demonstrated *in vitro* that MRI can be used as a tool for noninvasively assessing temperature changes in tissues associated with absorption of nonionizing radiation. As an example, we proposed the use of special NMR thermo-sensitive contrast reagents (NMR-TSCR) based on paramagnetic Ln complexes. This technique can diagnose conditions such as inflammatory reactions and cancer. The mechanism of action of NMR-TSCR is based on the substantial dependence of lanthanide-induced shifts (LIS) in NMR spectra on temperature.

Coordination compounds of lanthanides (Ln) with β-diketones have been considered as photoluminescence devices and photoresists as well as precursors for producing covers by the MO CVD method [1,2]. The pivaloyltrifluoracetonate ion, $(ptf)^-$, is among interesting ligands of β-diketone anions. The ligand involves fluorine atoms usable for NMR-sensors. Metal chelates with this ligand are thermostable and volatile. We previously researched by ^{1}H, ^{13}C, ^{19}F NMR (in solution) and by X-ray diffraction (in the crystal state) the starting compound $[Ln(H_2O)(ptf)_3]$ and the products formed by reacting with different ligands [3,4]. Furthermore, the temperature dependence of the signals of ^{19}F NMR of paramagnetic lanthanide complexes previously has rarely been studied in detail.

IOP Conf. Series: Journal of Physics: Conf. Series **886** (2017) 012003 doi:10.1088/1742-6596/886/1/012003

The aim of this work was ^{1}H and ^{19}F NMR investigation of $[Ln(PTA)_2(18\text{-crown-}6)]^+$ (Ln= La, Nd, Pr and Eu; Figure 1) and $[Ho(EDTA)]^+$ (Figure 6) as NMR thermo-sensitive contrast reagents (NMR-TSCR). We also studied the substantial temperature dependence of the paramagnetic lanthanide-induced shifts (LIS) in *tris*-pivaloyltrifluoracetonate lanthanides(III) (in $CDCl_3$ and $CD_3C_6D_5$; Figure 2 and 3) and $[Ho(EDTA)]^+$ (in D_2O; Figure 5) [3-5].

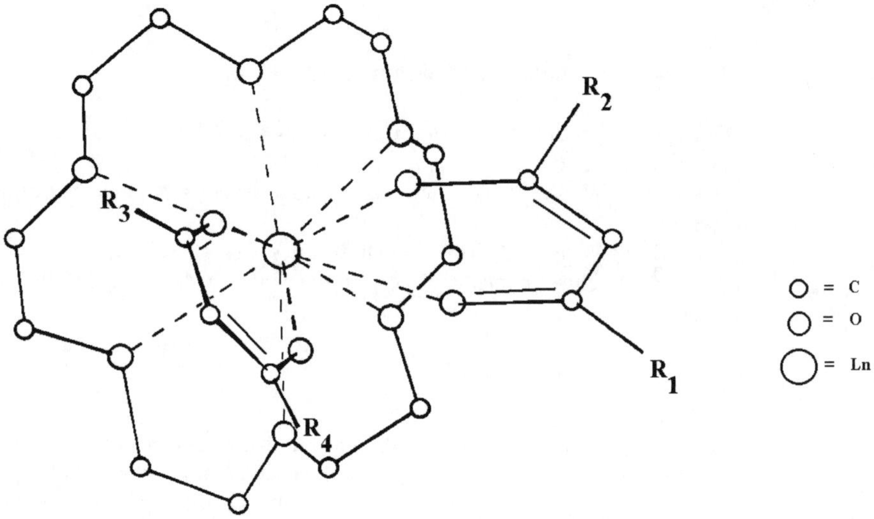

Figure 1. The proposed structure of the complex ions $[Ln(PTA)_2(18\text{-crown-}6)]^+$, where Ln= La, Ce, Pr, Nd; R_1= R_3= -Bu^t; R_2= R_4= $-CF_3$.

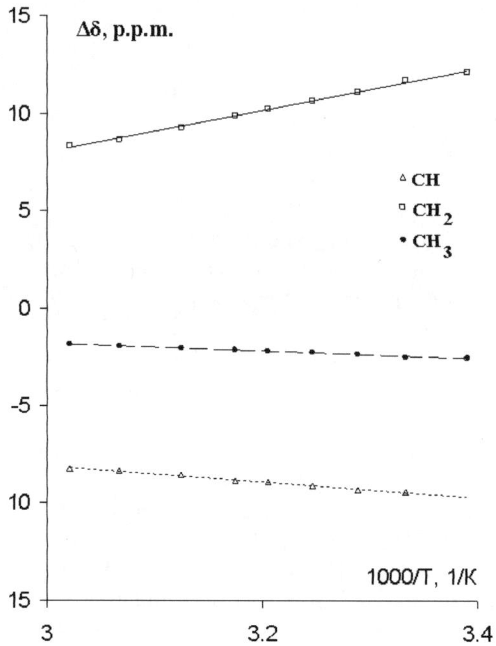

Figure 2. The temperature dependence of the paramagnetic lanthanide-induced shifts in the spectra of the 200 MHz ^{1}H NMR of the complex cation $[Eu(PTA)_2(18\text{-crown-}6)]^+$ in the solvent $CDCl_3$ (taken with respect to the corresponding protons of the diamagnetic complex cation $[La(PTA)_2(18\text{-crown-}6)]^+$).

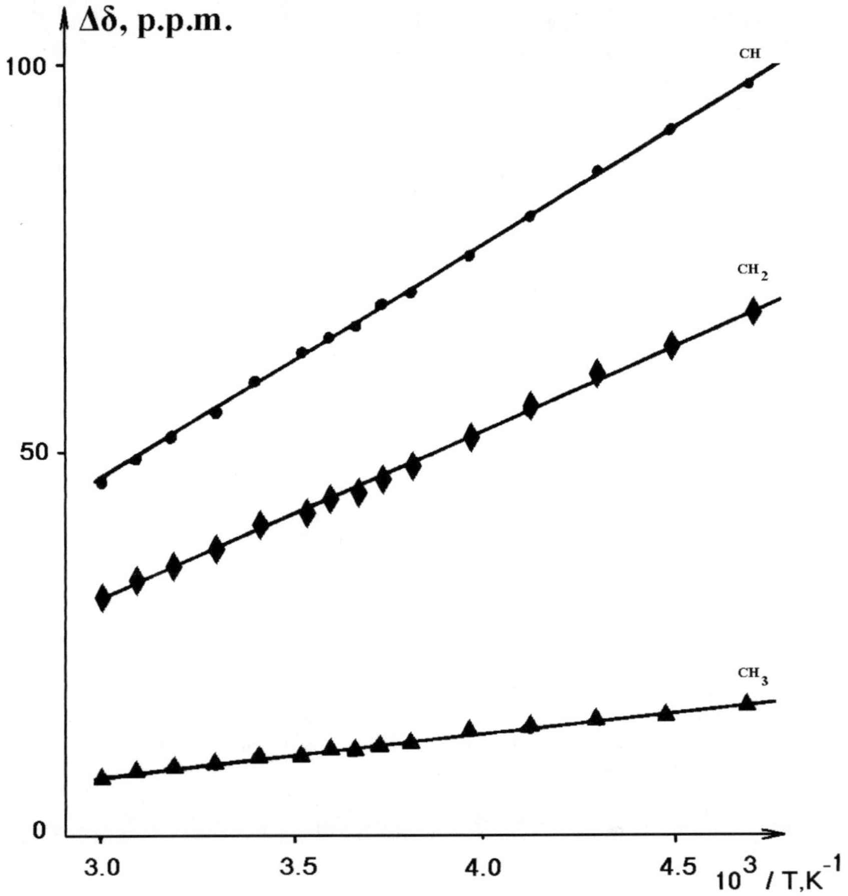

Figure 3. The temperature dependence of the paramagnetic lanthanide-induced shifts in the spectra of the 100 MHz ^1H NMR of the complex cation $[Pr(PTA)_2(18\text{-crown-6})]^+$ in the solvent $CDCl_3$ (taken with respect to the corresponding protons of the diamagnetic complex cation $[La(PTA)_2(18\text{-crown-6})]^+$).

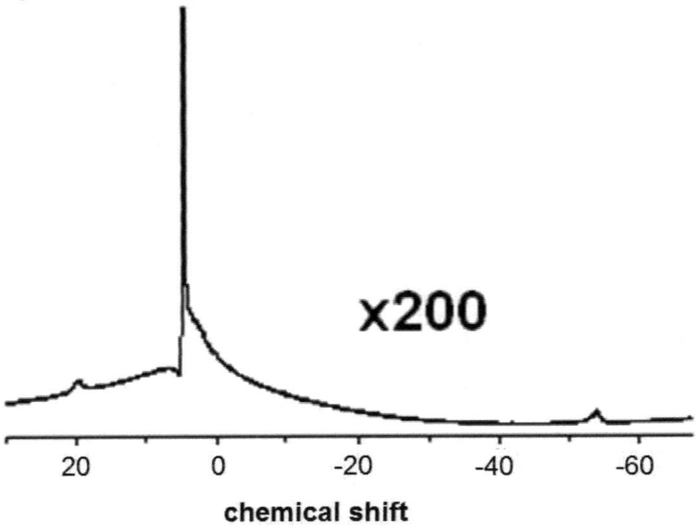

Figure 4. 300 MHz ^1H NMR of the complex $[Ho(EDTA)]^+$ in the solvent D_2O (used the sensor for microtomography, a diameter of 5 mm).

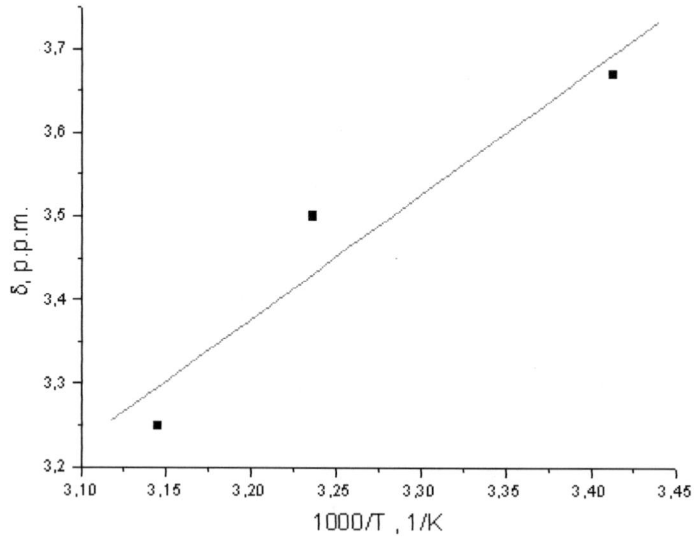

Figure 5. The temperature dependence of the paramagnetic lanthanide-induced shifts in the spectra of the 470.59 MHz ^{19}F NMR of the complex cation $[Pr(PTA)_2(18\text{-crown-}6)]^+$ in the solvent $CDCl_3$ (taken with respect to the corresponding protons of the diamagnetic complex cation $[La(PTA)_2(18\text{-crown-}6)]^+$); by BRUKER BioSpec 117/16 scanner.

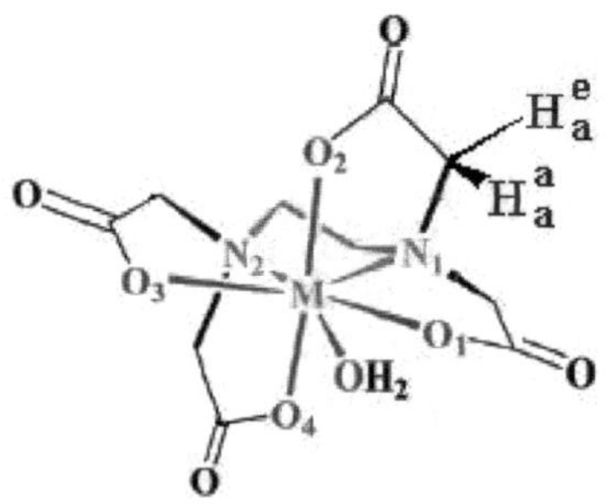

Figure 6. The proposed structure of the $[Ho(EDTA)]^+$ complex.

2. Results and discussion

It was found by ^{19}F NMR that the experimental paramagnetic LIS values (δ_{LIS}) of $[Ln(PTA)_2(18\text{-crown-}6)]^+$ {for the signals of CF_3} are well fitted by linear dependence on 1/T (Figure 5). This result correlates well with ^1H NMR studies of many other paramagnetic compounds of Ln cations.

We investigated experimental ^1H and ^{19}F NMR paramagnetic LISs versus T for the system in order to compare with analogous literature data. The derivative of δ_{LIS} (T) was calculated for the signals. These values appeared to be 0.021 for $[Pr(PTA)_2(18\text{-crown-}6)]^+$ (in case of ^{19}F NMR). The value of $\partial\delta/\partial T$ for the CH signal of $[Pr(PTA)_2(18\text{-crown-}6)]^+$ is 0.46 p.p.m./K (in case of ^1H NMR). The value of $\partial\delta/\partial T$ for the CH_2 signal of $[Ho(EDTA)]^+$ is 0.9 p.p.m./K (in case of ^1H NMR). It should be noted that other complex compounds, which were earlier investigated by ^1H NMR in both organic and aqueous media, are characterized by the LIS temperature sensitivity in the range from 0 to 1.17 p.p.m./K [4,6-12]. The obtained results indicate that the coordination compound $[Pr(PTA)_2(18\text{-crown-}6)]^+$ may be considered as sensitive ^1H and ^{19}F NMR lanthanide paramagnetic probe for *in situ* temperature control in organic solutions. It should be noted that this is a second determination of the temperature sensitivity of LIS in ^{19}F

NMR spectra of paramagnetic complexes of lanthanides. The compound $[Ho(EDTA)]^+$ may have a medical use for determining the local temperature of the body using MRI. The compound $[Pr(PTA)_2(18\text{-crown-6})]^+$ may also have a medical use for determining the local temperature of the body using MRI [7] in the case of application of coating technology (for example similarly to the publication [13]).

3. Experimental section

General. All reagents and solvents were commercially available and were used without additional purification. Elemental analysis was performed on a EuroEA 3000 analyzer. The melting points were determined on a Boëtius hot stage. The IR spectra were measured on a Scimitar FTS 2000 spectrometer in KBr in the range 375–4000 cm–1. Thermogravimetry was performed using a NETZSCH TG 209 F1 Iris thermomicrobalance (sample weight ~7 mg, aluminum crucible) in a helium atmosphere, gas flow rate 60 mL/min, heating rate 10 deg/min. The results were treated using Proteus Analysis software.

NMR spectroscopy. Solutions in $CDCl_3$ and CD_2Cl_2 had the concentration of complexes $C = 5{\cdot}10^{-3}$ M. The ^1H NMR spectra were measured with MSL-300, Avance-300, Avance-500 spectrometers and 500 MHz BioSpec 117/16 scanner (all produced by Bruker, Germany). The residual proton signal in $CDCl_3$ (7.23 p.p.m. at 298 K) was used for chemical shift referencing. Through all the measurements the magnetic field was stabilized, using deuterium lock of the solvent. The spectra were measured in the tubes with an outer diameter of 5 mm. A B-VT-1000 temperature unit with the accuracy of 0.5 K and stability of 0.2 Kh^{-1} was used for temperature control. The unit was calibrated, using standard samples with known temperature dependence of chemical shifts. Paramagnetic LISs were determined relative to the diamagnetic complex, which served as a reference. As it can be seen in Fig. 2, 3 and 5, experimental LIS values are well fitted, assuming their linear dependence on 1/T.

4. In conclusion

According to our evaluations, temperature determination error (ΔT) is ~0.4 K and 3D resolution (ΔL) is 100 mkm (see Fig. 7). To solve the main problem requires answers to a number of interrelated issues: (1) development of the method of the molecular structure determination of paramagnetic Ln complexes in solution according to the combined analysis of LIS and of the paramagnetic spin-lattice relaxation enhancement; (2) the information about the molecular structure and the dynamics of Ln coordination compounds with macrocyclic (crown-ethers) and acyclic (EDTA-like) ligands in solutions. The temperature dependencies of LISs revealed that Ln complexes might have practical importance for ^1H and ^{19}F NMR temperature control in solutions. It was shown *in vitro* (using NMR spectroscopy and MRI scanners) that these complexes can serve as NMR-TSCR and CAs.

Figure 7. The proposed image of temperature gradient at cancer and inflammation.

Acknowledgements

The work was fulfilled with the partial financial support of the Russian Foundation for Basic Research (Grants N14-03-00386-a, 18-53-53018 and 17-53-80070). Thanks to N.I. Anedchenko (Bogatireva) and

IOP Conf. Series: Journal of Physics: Conf. Series **886** (2017) 012003 doi:10.1088/1742-6596/886/1/012003

N.B. Egorov (National Research Tomsk Polytechnic University) for technical assistance. The publication was support by Russian Science Foundation, project No.14-45-00040-П.

References

[1] Malandrino G, Nigro R Lo, Fragala I L and Benelli C 2004 *Eur J Inorg Chem.* 500
[2] Katkova M A, Bochkarev M N 2010 *Dalton Trans.* **39** 6599
[3] (a) Babailov S P 2008 *Prog. Nucl. Magn. Reson. Spectrosc.* 1 1. (b) Voronov V K 2010 *Russ. Chem. Rev.* **79** 835. (c) Piguet C and Geraldes C F 2003 Handbook on the Physics and Chemistry of Rare Earths. *Elsevier Science* Amsterdam
[4] Babailov S P, Stabnikov P A, Korolkov I V, Pervukhina N V and Koshcheeva O S 2016 *Polyhedron* **105** 178 (PII: S0277-5387(15)00764-0; DOI: 10.1016/j.poly.2015.12.014)
[5] De Boer J W M, Sakkers P J D, Hilbers C W and De Boer M 1977 *J. Mag.Reson.* **25** 455
[6] Aime S, Barge A, Bruce,J I, Botta M, Howard J A K, Moloney J M, Parker D de Sousa A S and Woods M 1999 *J. Am. Chem. Soc.* **121** 5762
[7] Babailov S P, Dubovskii P V and Zapolotsky E N 2014 *Polyhedron* **79** 277 (DOI:10.1016/j.poly.2014.04.067)
[8] Babailov S P 2017 *Sensors & Actuators: B. Chemical* November **251** pp 108–111 (https://doi.org/10.1016/j.snb.2017.05.033)
[9] Babailov S P, Zapolotsky E N, Fomin E S and Qu Ya 2017 *Polyhedron* **134** pp 316-318 (http://www.sciencedirect.com/science/article/pii/S0277538717304035)
[10] Babailov S P 2016 *Sensors & Actuators: B. Chemical* B233 pp 476-478 (http://dx.doi.org/10.1016/j.snb.2016.04.009)
[11] Babailov S P, Peresypkina E V, Journaux Y and Vostrikova K E 2017 *Sensors & Actuators: B. Chemical* B 239 pp 405–412 (http://www.sciencedirect.com/science/article/pii/S0925400516312473?utm_campaign=49862_AUTH_UP_GC&utm_campaignPK=253284656&utm_term=OP26437&utm_content=254943554&utm_source=30&BID=753132795&utm_medium=email&SIS_ID=0)
[12] Babailov S P 2009 Progr. Nucl. Magn. Reson. Spectrosc. **54** № 3-4 p 183-194 (http://authors.elsevier.com/jpa/JPNMRS1262/8840b10885c171a9a6636d8bd3bdba76)
[13] Tanaka T, Mangala S, Vivas-Mejia P E, Nieves-Alicea R, Mann A P, Mora E, Han H D, Shahzad, M M K, Liu X, Bhavane R, Gu J, Fakhoury J R, Chiappini C, Lu C, Matsuo K, Godin B, Stone R L, Nick A M, Lopez-Berestein G, Sood A K and Ferrari M **2010** *Cancer Res.* **70** 3687 (PMID: 20430760; doi: 10.1158/0008-5472.CAN-09-3931)

Influence of magnetic field on enzymatic $ONOO^-$ production

T Dranova[1], D Petrovskii[1,2], N Ershov[2], I Slepneva[1] and D Stass[1]

[1]Voevodsky Institute of Chemical Kinetics and Combustion, Novosibirsk, Russia
[2]Federal Research Center Institute of Cytology and Genetics, Novosibirsk, Russia

E-mail: dm_petr@bionet.nsc.ru

Abstract. Enzymatic oxidation of L-arginine catalyzed by inducible nitric oxide synthase gives nitric oxide as the main product and superoxide anion as a side reaction product. Recombination of these radicals gives a very reactive species – peroxynitrite, which is involved in many biochemical processes. In the current work it was shown that such a system can be a usable model system for investigating the influence of magnetic field on enzymatic peroxynitrite formation. Using a selective fluorescent probe for peroxynitrite - coumarin boronic acid and an adopted for the experimental purpose incubation mixture, magnetic field experiments have been done at 11.7T. The averaged magnetic field effect is equal to 2.8±0.9%.

1. Introduction

In our previous work [1] we have shown that the recombination of nitric oxide (NO) and superoxide anion ($O_2^{\bullet-}$) that forms peroxinitrite ($ONOO^-$) is magnetosensitive. The radical pairs were formed by decomposition of 3-morpholinosydnonimine (SIN-1) in aqueous buffered solution at pH=7.6 (1).

$$SIN\text{-}1 \rightarrow NO + O_2^{\bullet-} \rightarrow ONOO^- \quad (1)$$

As described in numerous studies (see, e.g., [2]) magnetic field can affect spin dynamic in radical pairs and thus the yield of the spin-selective recombination product. The measured magnetic field (MF) effect in the system of SIN-1 in the field of 18 T equals 5.5±1.6% [1].

The purpose of current work is trying to detect this effect in a more complicated biochemical system, where the radical pairs are formed during enzymatic oxidation of L-arginine catalyzed by inducible nitric oxide synthase (iNOS) (2).

$$NOS \rightarrow NO + O_2^{\bullet-} \rightarrow ONOO^- \quad (2)$$

Nitric oxide synthases (NOS) are a family of enzymes catalyzing the production of NO from L-arginine (L-arg) [3] (3).

$$L\text{-}arg + 3/2\ NADPH + 1/2\ H^+ + 2\ O_2 \rightleftharpoons citrulline + NO + 2H_2O + 3/2\ NADP^+ \quad (3)$$

NO can be produced by nervous tissue (by neuronal NOS), by endothelium (endothelial NOS), and by macrophages (inducible NOS) [3]. For the purpose of this project inducible NOS was chosen, since iNOS is producing 100-1000 times more NO than other NOS isoforms. This allows NO detection at lower NOS concentration. This is important as NOS is thermally unstable and quite expensive.

NOS isoforms also catalyze other leak and side reactions, such as superoxide production. For all these isoforms superoxide production has also been detected [4-6]. So, in the catalytic cycle of NOS diffusion radical pairs $NO/O_2^{\bullet -}$ are generated.

2. Preliminary experiments

2.1. Optimization of incubation mixture

The main experimental challenge was to measure the yield of peroxynitrite formed as a result of enzymatic oxidation of L-arg. For this purpose the composition of iNOS incubation mixture was optimized, since most of the components are potential targets for peroxynitrite (including the buffer – HEPES [7]). Step by step these components (such as tetrahydrobiopterin BH4, dithiothreitol DTT, NADPH, and HEPES) were excluded or varied checking iNOS activity. The activity of NOS was detected by measuring the rate of NO formation. This was done by the hemoglobin method [8], which is based on the reaction of NO with oxyhemoglobine to form the oxidized form of hemoglobine (methhemoglobine). Oxyhemoglobine has a characteristic absorption band, so we can observe the decreasing optical density of oxygemoglobine at its characteristic maximum with time. The slope of this curve gives the rate of NO formation in the incubation mixture. It was estimated as 0.1-0.2 mkM/min. Checking the activity of iNOS at each step, it was decided to exclude from the incubation mixture BH4 and DTT, which is usually added to prevent oxidation of BH4. Also the concentration of NADPH was decreased and the concentration of Mg2+ was optimized in the new buffer (decreased) without a serious loss of NOS activity. HEPES buffer was changed to PBS. So, the new optimized incubation mixture for MF experiments on the NOS system was as follows: PBS (50 mM, pH=7.4), DTPA (50 mkM), MgCl2 (4.5 mM), iNOS (1-3 U/mL), L-arginine (1 mM), NADPH (50 mkM).

2.2. Peroxynitrite detection

In our work [1] peroxynitrite was detected using its reaction with dihydrorhodamine-123 DHR-123 and measuring the specifically colored product spectrophotometrically. This is not possible in the new system, since peroxynitrite readily reacts with NADPH ($k=4\times10^3\ M^{-1}s^{-1}$) [9] (rate constant for DHR-123 is $8.2\times10^3\ M^{-1}s^{-1}$ [10]). Therefore, for peroxynitrite detection coumarin boronic acid (CBA), a more specific probe, was used. CBA reacts with peroxynitrite with rate constant $k=1\times10^6\ M^{-1}s^{-1}$ forming hydroxycoumarin COH [11] that has a characteristic fluorescence band (λ_{excit}=322 nm, λ_{emit}=455 nm, Figure 1).

In order to estimate the yield of peroxynitrite formed during the enzymatic reaction, COH signal was calibrated as follows. To a known amount of CBA (25 mkM) an excess amount of SIN-1 (150 and 500 mkM) was added. The produced yield of COH, equal to 25 mkM, gives the fluorescence intensity reaching 700 mV. During the enzymatic oxidation the COH signal was rising up to 150 mV. So, the estimated yield of peroxynitrite is about 5 mkM. Taking into account the fact that the error of fluorescence signal measurements is about 0.5%, the yield of peroxynitrite should be enough to see a MF effect about 2-3% that is expected in this system.

Using the CBA probe the effect of NOS concentration on the rate of COH formation was investigated. It was found that NOS concentration affects the initial rate of peroxynitrite formation and also the running time of peroxynitrite formation. Beyond a certain NOS concentration a plateau was reached. This fact can be explained by autoinhibition effect. So, for better experimental conditions, the concentration of NOS of 1-3 U/mL was chosen. This concentration is enough to produce peroxynitrite yield with good fluorescent signal of COH (not being at the plateau). The running time at this concentration is about 1.5 hours.

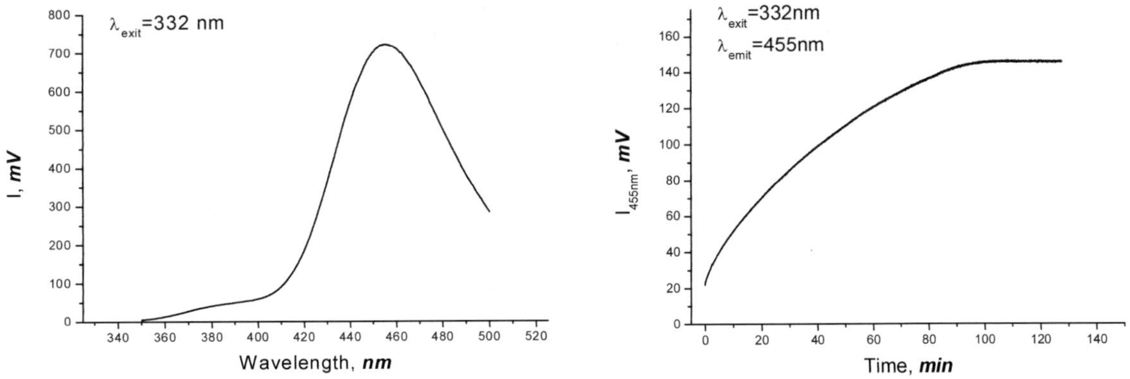

Figure 1. Left panel: fluorescence spectrum of COH, the product of CBA oxidation (λ_{excit}=322 nm, λ_{emit}=455 nm). Right panel: kinetics of COH formation in NOS incubation mixture during CBA oxidation by peroxinitrite (PBS (50 mM, pH=7.4), DTPA (50 mkM), $MgCl_2$ (4.5 mM), iNOS (1.5 U/mL), L-arginine (1 mM), NADPH (50 mkM), CBA (25 mkM). T=T_{room}.

3. Materials and methods

Reagents: Standard phosphate buffer solutions of pH 7.6 were prepared from deionized water, K_2HPO_4, and KH_2PO_4 (99.0 %, Fluka). To eliminate traces of transition metal ions, diethylenetriamine pentaacetic acid (DTPA, 99.0 %, Fluka) was added at a concentration of 50 mkM. Inducible NOS (Cayman Chemicals) was stored at -70 ºC, Stock 1 mM solutions of SIN-1 (3-morpholinosydnonimine hydrochloride, Invitrogen) in pH 7.6 buffer was also stored at -20 ºC after preparation. Oxyhemoglobine was prepared from Hemoglobine (Sigma) as described in ref. [8]. For quantifying oxyhemoglobine characteristic absorption maxima (ε_{577nm}=15.4 $mM^{-1}cm^{-1}$) was used [8]. CBA (Cayman Chemicals) 1 mM stock solution in DMSO was stored at -20 ºC.

The absorption spectra were taken on a Shimadzu UV-2401 UV/Vis spectrometer. The fluorescent measurements were carried out on a Hitachi MPF4 fluorimeter or Perkin Elmer Victor X3 2030 Multilabel Reader.

4. Magnetic field experiments

MF experiments were performed in the magnet of a high field MRI scanner (11.7 T) Bruker BioSpec 117/16. To control the temperature two identical thermostats were built. One was placed inside the magnet, and the other was left in the room for the control sample. Temperature stabilization was achieved by pumping water through the two thermostats connected in parallel. The thermostats were checked before the experiments, the temperature inside them was equal to within 0.2º C.

MF experiments were carried out using two groups of samples (for incubation in the magnet and for control) that were prepared from the same incubation mixture. Each group comprised 8 samples (16 altogether). Incubation time was 3 hours.

For the fluorescence measurements in MF experiments a multi-channel fluorimeter was used, which can measure all the samples (one by one at the fluorescence maximum) in a very short time. One run of such measurements takes only about 10 sec. For reliability, the samples were placed at different positions over the plate (a type of plate formed by individual strips) and then the results were averaged. The volume of each sample was 200 mkL.

Magnetic field effect was calculated as $\varphi = \dfrac{\bar{I}_m - \bar{I}_c}{\bar{I}_c}$, where $\bar{I}_m$ and $\bar{I}_c$ are averaged over eight samples fluorescence intensities for samples incubated in magnetic field and controls, respectively. Ten MF experiments have been done in the field 11.7 T (Figure 2 (a)). The error of the measured effect was calculated as the standard error of the mean $S_{\bar{\varphi}} = \frac{S}{\sqrt{n}}$, where n is the number of experiments in the field 11.7 T. Several otherwise identical experiments in the bore of the magnet that was quenched have also been done. The results are shown in Figure 2 (b). Significance of the magnetic field effect was estimated by two-way ANOVA with factors "magnet" and "day of experiment".

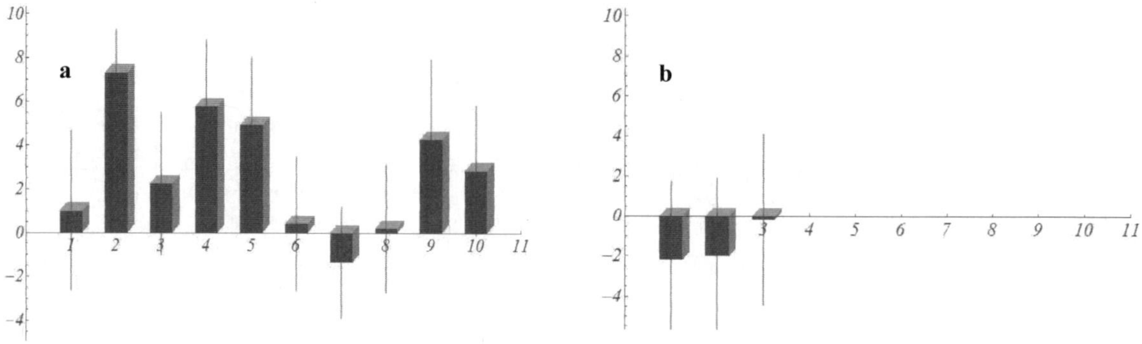

Figure 2. (a) Results of 10 MF experiments: (1.0±3.7)%, (7.3±2.0)%, (2.3±3.3)%, (5.8±3.0)%, (4.9±3.0)%, (0.4±3.1)%, (-1.3±2.6)%, (0.2±2.9)%, (4.3±3.7)%, (2.9±3.0)%. The averaged magnetic field effect is equal to (2.8±0.9)%. **(b)** Results of zero field experiments: (-2.2±4.0)%, (-2.0±3.9)%, (-0.2±4.3)%. The averaged effect is equal to (-1.5±0.6)%.

5. Results and Discussion

In continuation of the work [1] experiments to search for the magnetic field effect on the more complex biochemical system of enzymatic oxidation of L-arginine catalyzed by inducible NO-synthase were prepared and carried out. As a result of this reaction, as in the case of decomposition of 3-morpholinosydnonimine, radical pairs NO and $O_2^{\bullet -}$ are formed, promising to give a magnetically sensitive recombination to form peroxynitrite. The expected magnetic effect is 2-3% of the yield of peroxynitrite.

In order to make possible measurements of the yield of peroxynitrite, the composition of iNOS incubation mixture was optimized. Potential targets for peroxynitrite were excluded or varied without a serious loss of NOS activity. For detection of peroxynitrite a selective fluorescent probe - coumarin boronic acid (CBA) was used.

In the current work ten magnetic field experiments have been done at 11.7 T. The averaged magnetic field effect has been calculated, which is equal to 2.8±0.9% ($F_{1,142}$=35.9, $p<0.001$). Control experiments at zero field give the result -1.5±0.6% ($F_{1,42}$=2.7, $p>0.1$). This may indicate a probable magnetic field sensitive step in the enzymatic formation of peroxynitrite.

Acknowledgments
The authors acknowledge the support by the Russian Science Foundation, grant No. 14-45-00040-П for the manuscript preparation and publication; the Center for Genetic Resources of Laboratory Animals at the Institute of Cytology and Genetics for the opportunity to use the MRI scanner, Integration project No. 71 and budget project No. 0324-2016-0002 of the Federal Research Center of the Institute of Cytology and Genetics, SB RAS; as well as Bobko A A, Pomaznoy M Yu, Dranov I G for significant project support.

IOP Conf. Series: Journal of Physics: Conf. Series **886** (2017) 012004 doi:10.1088/1742-6596/886/1/012004

References

[1] Karogodina T Yu, Dranov I G, Sergeeva S V, Stass D V, Steiner U E 2011 Chem. Phys. Chem. **12** 1714–28
[2] Steiner U E, Ulrich T 1989 Chem. Rev. **89** 51–147
[3] Griffith O W, Stuehr D J 1995 Annu. Rev. Physiol. **57** 707–36
[4] Xia Y, Roman L J, Masters B S, Zweier J L 1998 J. Biol. Chem. **273** 22635–39
[5] Pou S, Pou W S, Bredt D S, Snyder S H, Rosen G 1992 J. Biol. Chem. **267** 24173–76
[6] Vásquez-Vivar J, Kalyanaraman B, Martásek P, Hogg N, Masters B S, Karoui H, Tordo P, Pritchard K A Jr. 1998 Proc. Natl. Acad. Sci. USA **95** 9220–25
[7] Kirsch M, Lomonosova E E, Korth H, Sustmann R, Groot H 1998 J. Biol. Chem. **273** 12716-24
[8] Feelisch M, Kubitzek D, Werringloer J 1996 *Methods in Nitric Oxide Research* ed M Feelisch and J Stamler (Chichester, Wiley) pp 455–78
[9] Kobayashi K, Miki M, Tagawa S 1995 Endothelium **3** (suppl), 65.
[10] Kumar S S, Priyadarsini K I, Sainis K B 2004 J. Agric. Food. Chem. **52** 139–145.
[11] Zielonka J, Sikora A, Joseph J, Kalyanaraman B 2010 J. Biol. Chem. **285**, 14210–16

IOP Conf. Series: Journal of Physics: Conf. Series **886** (2017) 012005 doi:10.1088/1742-6596/886/1/012005

Effects of insulin-like growth factor 1 on pathologic processes in the cuprizone model of multiple sclerosis

D Fedorishin[1], I Sorokina[2], T Tolstikova[2], A Akulov[1,3], V Glazacheva[1], N Nemirovich-Danchenko[1], M Khodanovich[1] and V Yarnykh[1,4]

[1] Research Institute of Biology and Biophysics, Tomsk State University, Tomsk, Russia
[2]Vorozhtsov Novosibirsk Institute of Organic Chemistry, Novosibirsk, Russia
[3]Institute of Cytology and Genetics of the Siberian Branch of the Russian Academy of Sciences, Novosibirsk, Russia
[4]Vascular Imaging Laboratory, Department of Radiology, University of Washington, Seattle, WA, USA

E-mail: khodanovich@mail.tsu.ru

Abstract. The study aims to evaluate the effect of insulin-like growth factor 1 (IGF-1) on the demyelination and astrogliosis using the cuprizone murine model. Demyelination was induced in 14 adult male mice by 0.3% cuprizone in drinking water. Five animals from the cuprizone-treated group received subcutaneous injections of IGF-1. Seven animals were used as a control group. The extent of demyelination was evaluated as a decrease in the size of the corpus callosum on T2-weighted images that were received using an 11.7T animal MRI scanner. Brain sections were immunohistochemically stained for glial fibrillary acidic protein (GFAP), a marker of astrocytes. It was revealed that the cuprizone caused extensive demyelination and astroglyosis. IGF-1 treatment restored the size of the corpus callosum and the number of astrocytes in the corpus callosum and the anterior commissure to the control level.

1. Introduction

Multiple sclerosis (MS) is one of the most common demyelinating diseases. Drugs whose action is aimed to decrease demyelination and stimulate remyelination processes still remain mostly at the stage of experimental testing [1,2]. The cuprizone administration induces extensive myelin damage both in white and gray matter [3-5], which can be evaluated by MRI [6-8] and histologically [3-7]. Oligodendrocytes are a primary type of cells affected by cuprizone [3-5]. However, the other types of neural cells including neurons [2,9,10] and astrocytes [3,11,12] are also affected by cuprizone administration even in the absence of autoimmune processes, though to a lesser extent. Insulin-like growth factor 1 (IGF-1) is a perspective agent which can stimulate remyelination due to its influence on cell differentiation into oligodendrocytes [13]. The purpose of the research was to study the prospects of using IGF-1 for the reduction of the severity of pathological processes in the brain inflicted by the administration of cuprizone.

2. Materials and methods

The study involved three groups of male mice of C57Bl/6j line, 8 weeks of age, obtained from N.N. Vorozhtsov Novosibirsk Institute of Organic Chemistry (Novosibirsk): "Control" (N=7),

"Demyelination" (N=5) and "IGF-1" (N=7). Demyelination in the groups "Demyelination" and "IGF-1" was caused by a chronic oral administration of 0.3% aqueous solution of cuprizone for 53 days. The animals from the "IGF-1" group were injected with IGF-1 subcutaneously twice a week, and the animals from the other groups received injections of saline. On day 53, being under 1.5-2% isoflurane anesthesia, the animals underwent MRI scanning of the brain with the use of a BioSpec 117/16 USR ultra-high-angle tomograph (Bruker, Germany), which allowed obtaining T2-weighted images. The animals were then euthanized under ether anesthesia by transcardial perfusion with 4% paraformaldehyde. The brain was removed, and cryoprotection was performed in 10% and 20% solutions of sucrose, followed with freezing in liquid nitrogen vapor. For the assessment of astrogliosis, 10-μm-thick sections of the brain were obtained on the cryotome HM525 (TermoScientific, Germany) in the area of +0.38 mm from bregma, and then they were stained immunohistochemically with antibodies with an affinity for glial fibrillary acid protein (GFAP) – an astrocyte marker. Photomicrographs of sections taken with the AxioImager Z2 microscope (Carl Zeiss, Germany) and T2-weighted images were processed with ImageJ software. The corpus callosum was manually contoured on the T2-weighted images and its area was counted. In order to calculate the number of astrocytes, the areas of the corpus callosum of 200x200 μm were selected, as well as the front commissures within the natural boundaries. Statistical processing was carried out using Statistica 8.0 software with the help of variance analysis and the Mann-Whitney test.

3. Results

The T2-weighted images showed a significant decrease in the corpus callosum size in the "Demyelination" group as compared to the control group. Statistical analysis revealed a significant decrease in the size of the corpus callosum caused by the administration of cuprizone as compared to the control group ($p<0.001$). The administration of IGF-1 on the background of cuprizone intoxication restored the size of the corpus callosum to the control level ($p>0.05$).

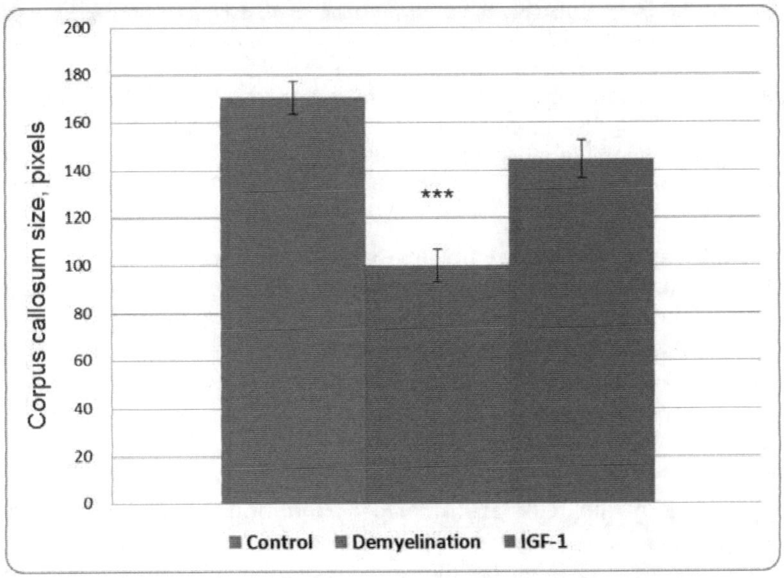

Figure 1. The size of the corpus callosum in the groups "Control", "Demyelination" and "IGF-1". The columns reflect the mean ± standard error of the mean. The significant differences compared to the "Control" group: *** – $p < 0.001$.

In the studied structures of the brain of the animals from the "Demyelination" group, a large number of hypertrophied astrocytes with a large number of processes was found, which indicates a pronounced intoxication with cuprizone. Fewer hypertrophied astrocytes were observed in the "IGF-1" group than in the "Demyelination" group. Statistical analysis revealed a significant increase in the number of astrocytes in the "Demyelination" group as compared to the control group both in commissures ($p<0.01$) and in the corpus callosum ($p<0.01$). At the same time, this indicator did not differ from the control level for the "IGF-1" group ($p>0.05$).

It is known that IGF-1 has anti-inflammatory properties, and can increase the survival and proliferation of oligodendrocytes, contributing to remyelination [2]. Presumably, a decrease in the number of active astrocytes on the background of the administration of cuprizone, which causes astrogliosis, is associated with the neuroprotective and remyelinating effect of IGF-1.

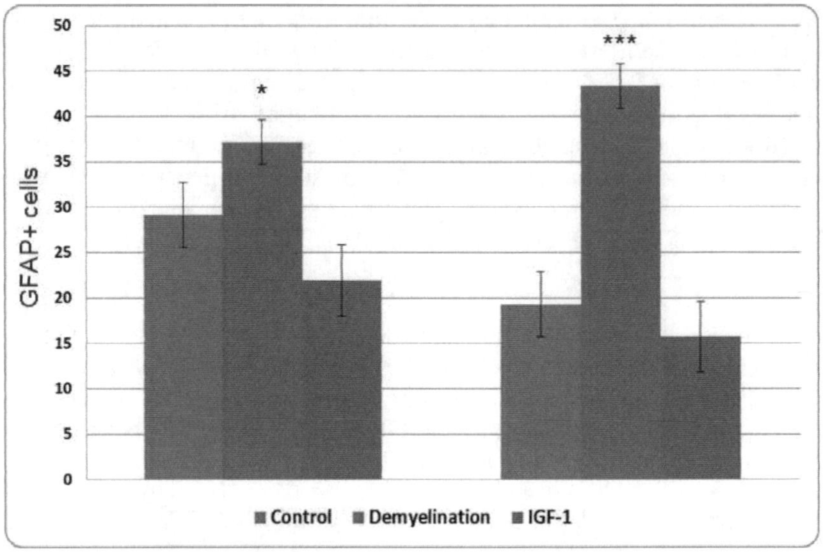

Figure 2. Differences in the number of astrocytes in the corpus callosum in the groups "Control", "Demyelination" and "IGF-1". The columns reflect the mean ± standard error of the mean. The significance of differences in relation to the "Control" group is * – $p < 0.05$; *** – $p < 0.001$.

4. Conclusion

A chronic cuprizone intoxication caused demyelination in mice, which resulted in a decrease in the size of the corpus callosum, an increase in the number of astrocytes in the corpus callosum and anterior commissure, and a change in their morphology. The administration of IGF-1 on the background of cuprizone intoxication had a positive effect on demyelination, increasing the size of the corpus callosum to control values and reducing the astrogliosis caused by cuprizone in the corpus callosum and in anterior commissures.

Acknowledgements

The study was supported by the Russian Science Foundation, project No.14-45-00040-П. The author expresses his gratitude to the staff members of Vorozhtsov Novosibirsk Institute of Organic Chemistry, I Sorokina and T Tolstikova, for the implementation of the cuprizone model of multiple sclerosis, to the staff members of the Institute of Cytology and Genetics of the Siberian Branch of the Russian Academy of Sciences, A Akulov and A Romashchenko, for conducting the MRI studies, and to the employees of the Neurobiology Laboratory of the Research Institute of Biology and Biophysics at Tomsk State University, N Nemirovich-Danchenko and V Glazacheva, for their help in the carrying out of the research.

References

IOP Conf. Series: Journal of Physics: Conf. Series **886** (2017) 012005 doi:10.1088/1742-6596/886/1/012005

[1] Virley D. 2005 Developing Therapeutics for the Treatment of Multiple Sclerosis *NeuroRx* **2(4)** 638–649

[2] Khodanovich M, Glazacheva V, Pan E, Akulov A, Krutenkova E,Trusov V and Yarnykh V.. 2015 MRI study of the cuprizone-induced mouse model of multiple sclerosis: demyelination is not found after co-treatment with polyprenols (long-chain isoprenoid alcohols). *Journal of Physics Conference Series* **677(2)** UNSP 012007.

[3] Kipp M, Clarner T, Dang J and Beyer C 2009 The cuprizone animal model: new insights into an old story *Acta Neuropathologica.* **6** 723

[4] Torkildsen O, Brunborg LA, Myhr K-M and Bo L 2008 The cuprizone model for demyelination Acta Neurol Scand 11 72.

[5] Acs P, Kalman B 2012 Pathogenesis of multiple sclerosis: what can we learn from the cuprizone model. *Methods Mol Biol* **900** 403–431.

[6] Wood T, Simmons C, Hurley S,. Vernon A, Torres J, Dell'Acqua F,. Williams S and Cash D. 2016 Whole-brain *ex-vivo* quantitative MRI of the cuprizone mouse. *PeerJ* **4** e2632.

[7] Khodanovich M, Sorokina I, Glazacheva V, Akulov A, Nemirovich-Danchenko N, Romashchenko A, Tolstikova T, Mustafina L and Yarnykh V 2017 Histological validation of fast macromolecular proton fraction mapping as a quantitative myelin imaging method in the cuprizone demyelination model. *Sci Rep.* **7** 46686

[8] Krutenkova E., Pan E., Khodanovich M. A 2015 Review of MRI Evaluation of Demyelination in Cuprizone Murine Model. *AIP Conference Proceedings* **1688** 030003.

[9] Hillis J, Davies J, Mundim M, Al-Dalahmah O and Szele F. 2016 Cuprizone demyelination induces a unique inflammatory response in the subventricular zone. *Journal of Neuroinflammation* **13(1)** 190

[10] Benetti F, Ventura M, Salmini B, Ceola S, Carbonera D, Mammi S, Zitolo A, D'Angelo P, Urso E, Maffia M, Salvato B, Spisni E 2010 Cuprizone neurotoxicity, copper deficiency and neurodegeneration *Neurotoxicology* **31** 509–517

[11] Minagar A, Shapshak P, Fujimura R [et al.] 2002 The role of macrophage/microglia and astrocytes in the pathogenesis of three neurologic disorders: HIV-associated dementia, Alzheimer disease, and multiple sclerosis *Neurol. Sci* **202** 13–23.

[12] Hatten M , Liem K, Shelanski M et al 1991 Astroglia in CNS injury *Glia* **4(2)** 233–243.

[13] Hsieh J, Aimone J, Kaspar B [et al.] 2004 IGF-I instructs multipotent adult neural progenitor cells to become oligodendrocytes J Cell Biol **164** 111–122.

Multinuclear magnetic resonance imaging in biomedical experiments

N Anisimov[1], M Gulyaev[1], O Pavlova[2], D Volkov[2], L Gervits[3] and Yu Pirogov[2]

[1]Faculty of Fundamental Medicine, Lomonosov Moscow State University, Moscow, Russia
[2]Faculty of Physics, Lomonosov Moscow State University, Moscow, Russia
[3]A. N. Nesmeyanov Institute of Organoelement Compounds, Russian Academy of Sciences, Moscow, Russia

E-mail: yupi937@gmail.com

Abstract. The detection of nuclei other than protons can significantly enhance the opportunities of MRI method. Thereby one of the most promising nuclei in this field is fluorine-19. We showed that it is possible to visualize lungs, gastrointestinal tract and inflammation regions by 19F MRI without complicated and expensive specific technics. Moreover, the use of other nuclei in multinuclear applications is also possible and it is a promising direction in this field.

1. Introduction. Purpose

Magnetic resonance imaging (MRI) is usually used for visualization of internal organs to elicit its pathologies, which appear as variances in geometrical proportions (morphological changes) and tissues composition. Last is elicited on MRI images by contrast changes. The contrast control in MRI is provided with setting parameters of impulse sequences, choice of which allows to improve imaging tissues with short or long relaxation times. To solve these challenges, MRI method on hydrogen nucleus (proton) – detection of NMR signals from hydrogen-contained compounds (water, fat, etc.) – is used. This method is very sensitive to protons due to its high gyromagnetic ratio and 100% nature abundance.

2. Materials and methods

The additional information about internal organs can be also obtained from other magnetic nuclei, which are contained in organic molecules. However, they have much less sensitivities because of their small gyromagnetic ratios. So, only the detection of phosphorus 31P and sodium 23Na (beside 1H) nuclei is applied in up-to-date MRI. The detection of such nuclei as 13C, 15N, 17O, 2D, etc. is usually applied for studying metabolic processes, whose labels are isotope-enriched compounds. The preliminary hyperpolarization of these compounds is applied to achieve the maximal sensitivity.

The typical configuration of medical MRI scanner is aimed to the detection only protons. However, in some cases, its hardware may be adjusted to detect nuclei other than protons. We described in our previous studies [1-4] how 0.5T medical MR scanner Bruker Tomikon S50 has been adjusted for detection not only protons, but also 19F, 31P, 13C, 23Na, 11B, 2H, 35Cl nuclei. Herewith 19F NMR spectroscopy and MRI were applied in *in vivo* studies with laboratory animals – rats, mice. The studies of the perfluorocarbon compound Perftoranum® (well-known as a blood substitute with efficient gas transmission function) were performed.

3. Results

With use of the 0.5T MRI scanner, we were able to detect the 19F MRI signals in the study of human gastrointestinal tract. The figure 1 a, b shows 19F and 1H MRI images of healthy volunteer, who has swallowed the capsule filled with perfluorodecalin (PFD). These images were obtained using 3D-gradient echo (GE). Scanning parameters for 19F MRI were followed: FA=300, TR/TE=600/4.6 ms, BWT/BWR=0.7/125 kHz, EXPT=20 min, MTX=128×128×16, FOV=30×30×35 cm.

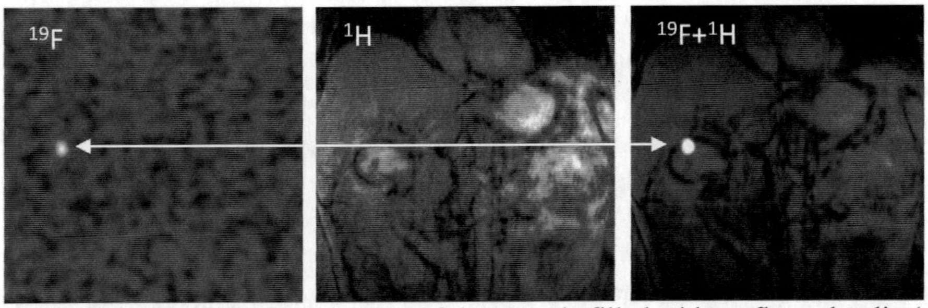

Figure 1. 19F-, 1H- and (19F+1H)-MRI images of the capsule filled with perfluorodecalin (as marked by the arrow) in gastrointestinal tract.

The combination of 1H and 19F MRI images allows to specify the localization of the capsule with PFD relatively to the anatomical structures. One hour after swallowing, this capsule was found (as it's shown in the Figure 1) in the stomach, at the entrance to the duodenum. Capsule shell with 1-ml internal volume was made from gelatin and polyacryl.

Scientific research MRI scanners intended for small laboratory animals (rats, mice, etc.) investigations are more suitable for multinuclear experiments. One of such scanner is 7T MRI device Bruker BioSpec 70/30 USR, in which we have studied metabolic processes using 13C and 31P NMR spectrometry. Acceptable results were also obtained by the detection of deuterium 2H with help of NMR spectroscopy and MRI methods [5].

An important direction in multinuclear MRI is lung visualization. Hyperpolarized gases, whose atoms are polarized to almost 100% condition by specific methods, are usually used for this purpose. The hyperpolarization condition is nonequilibrium and has practical interest only in the case of weakly interacting atoms, whose translations in the equilibrium condition occur for a long time. As such substances, the inert gases (3He, 199Xe, 36Kr, etc.) and gases, which contain nuclei with low gyromagnetic ratio (for example, isotope-enriched by 13C carbon dioxide) can be.

Because preparation of the hyperpolarized gases is technically complicated and consequently available to only few number of laboratory in the world, the fluorine-contained compounds can be used for this purpose as an alternative to hyperpolarized gases.

The detection of 19F NMR signal is promising method because its sensitivity is comparable to proton sensitivity. Herewith in the case of 19F MRI, there is no background from normal tissues on fluorine images, because the extremely little fluorine amount is contained in these tissues. Also because of the same reason, it is enough to apply only one impulse sequence – not two, three or more as it is in the case of 1H MRI – to differentiate the tissues by measuring 19F signals.

Since NMR resonance frequencies of protons and fluorin-19 are very close, it is possible to use the same technical hardware (receiver-transmitter, NMR signal detector) to detect these nuclei. For 0.5T MRI scanner, we used proprietary receive coils, whose tuning range was enough to work with both of the nuclei (19F, 1H). However proprietary transmitting coil could work only on protons. Therefore one of the quadrature channels of the receive coil was adjusted to transmit RF power.

For 7T MRI scanner, the proprietary resonator designed to detect only hydrogen was adjusted for detecting the fluorine-19 nuclei. Because the additional capacity and commutator were installed in this resonator circuitry, the own resonator's frequency could be switched depending on the fact which nuclei (hydrogen or fluorine) are needed to detect.

The interesting results were obtained using 19F frequency resonators with inductively (i.e. by wireless way) included coils in the program of creating so-called implant systems. The main purpose of such implant (the plane one-loop coil fine-tuned to the Larmor frequency) is to transmit and receive NMR signals at the region which is very close to the required specific organ. Thereby the level of RF field strength in the region can be significantly enhanced in the implant region reaching the best sensitivity [6]. Wireless coils were used in our studies to detect 19F MRI signals, but it is absolutely clear that such coils can be useful for other multinuclear applications.

The results (Figure 2) which show the effectiveness of fluorocarbon compounds as the contrast agent and different markers for imaging separate internal organs, gastrointestinal tract, lungs and inflammation regions were obtained by 19F MRI method in the *in vivo* studies of perfluorocarbon compounds injected to the body of laboratory animals. The rats "Wistar" as laboratory animals were used in these experiments.

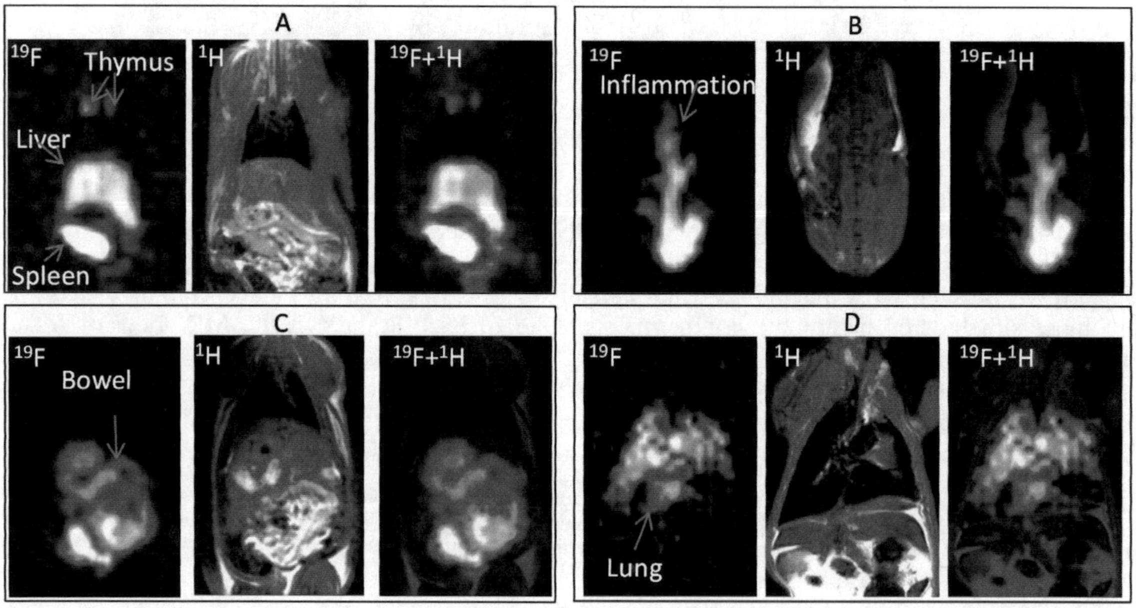

Figure 2. A - liver, spleen and thymus of rat after intraperitoneal injection of Perftoranum®; B- visualization of cellulitis of rat after intravenous injection of perfluorotributylamine (C12F27N); C - gastrointestinal tract of rat after oral administration of Perftoranum®; D - Lung of rat after inhalation of perfluorocyclobutane (C4F8).

A number of these obtained data are not obvious. Note, the concentration of Perftoranum® after its intraperitoneal injection became apparent not only in liver and spleen, but also in thymus as it is seen on the fragment A of the Figure 2. However, after intravenously injection, the Perftoranum® accumulates only in liver and spleen.

4. Discussion. Conclusions

Detection of nuclei other than protons can significantly expand opportunities of MRI method. Thereby, especially signuficant prospects relate to the application of 19F MRI. However the use of other nuclei in multinuclear applications presents also promising direction in this line. Such applications become possible using relatively simple technical resources (for example – implant coils) without isotope enrichment of sounding compounds or their hyperpolarization.

Acknowledgements
This work was supported by the grant from the Ministry of Education and Science of the Russian Federation No. 14.604.21.0060 (RFMEFI60414X0060).

References

[1] Volkov D V, Pavlova O S, Gulyaev M V, Anisimov N V and Pirogov Yu A 2014 *Journal of radio electronics* **11** - http://jre.cplire.ru/jre/nov14/7/text.pdf (*in Russian*)
[2] Anisimov N V, Pavlova O S, Volkov D V, Gulyaev M V and Pirogov Yu A 2015 *Journal of radio electronics* **11** - http://jre.cplire.ru/jre/nov15/12/text.pdf (*in Russian*)
[3] Anisimov N V, Volkov D V, Gulyaev M V, Pavlova O S and Pirogov Yu A 2016 *IOPscience Journal of Physics: Conference Series* **677** 012005
[4] Pirogov Yu A 2016 *Physics Procedia* **82C**
[5] Kosenkov A V, Gulyaev M V, Anisimov N V, Lobyshev V I, Pirogov Yu A 2015 *Physics of Wave Phenomena* **23(4)** 311–315
[6] Anisimov N V, Gulyaev M V, Volkov D V, Pavlova O S and Pirogov Yu A 2015 *Journal of radio electronics* **11** - http://jre.cplire.ru/jre/nov16/13/text.pdf (*in Russian*)

IOP Conf. Series: Journal of Physics: Conf. Series **886** (2017) 012007 doi:10.1088/1742-6596/886/1/012007

Resting-state fMRI study of patients with fragile X syndrome

E Isanova[1], E Petrovskiy[3], A Savelov[2,3], D Yudkin[1,2] and A Tulupov[1,2,3]

[1] Institute of Medicine and Psychology, Novosibirsk State University, Novosibirsk, Russia,
[2] Institute of Molecular and Cellular Biology of the Russian Academy of Sciences, Novosibirsk, Russia,
[3] Institute International Tomography Centre of the Russian Academy of Sciences, Russia,

E-mail: ise_1993@mail.ru

Abstract. The study aimed to assess the neural activity of different brain regions in patients with fragile X syndrome (FXS) and the healthy volunteers by resting-state functional magnetic resonance imaging (fMRI) on a 1.5 T MRI Achieva scanner (Philips). Results: The fMRI study showed a DMN of brain function in patients with FXS, as well as in the healthy volunteers. Furthermore, it was found that a default mode network of the brain in patients with FXS and healthy volunteers does not have statistically significant differences ($p>0.05$), which may indicate that the basal activity of neurons in patients with FXS is not reduced. In addition, we have found a significant ($p<0.001$) increase in the FC within the right inferior parietal and right angular gyrus in the resting state in patients with FXS. Conclusion: New data of functional status of the brain in patients with FXS were received. The significant increase in the resting state functional connectivity within the right inferior parietal and right angular gyrus ($p<0.001$) in patients with FXS was found.

1. Introduction

Fragile X syndrome (FXS) is one of the most common causes of hereditary mental retardation. The main mechanism of pathogenesis associated with the expansion CGG repeats in the promoter region of the FMR1 (fragile mental retardation 1) gene, resulting in suppression of production of protein FMRP (fragile mental retardation protein), which is essential in the normal development of the central nervous system [1]. The importance of early diagnosis is determined by the increasing severity of the disease from generation to generation, which is associated with the accumulation of the number of repeats in the mutant site the of the X-chromosome. During the last decade, researchers have been actively using resting-state (fMRI) for the study of brain physiology and pathological changes. This technology allows practitioners to evaluate the low-frequency (0.01–0.1 Hz) fluctuations in BOLD signal, and it is most frequently used to study patients with neurological or psychiatric disorders. BOLD signal is based on changes in hemodynamic parameters, and it indirectly detects the neuronal activity of the various brain regions both during task completion and in resting state. Resting-state fMRI allows the assessment of functional connectivity and the identification of the default mode network (DMN). In contrast to other networks, DMN has elevated levels of neuronal activity during rest, compared to when tasks are performed. It is accepted that the activities of the network show the basal activity of neurons of the human brain.

In the present study, we used a new approach to the diagnosis of diseases, such as FXS, based on neuroimaging techniques. This gives hope that the data obtained from the study of this disease can help in the search of mental retardation mechanisms. Furthermore, the data obtained from the study areused as a prediction of mental retardation and to select strategies of cognitive therapy. Usually, when performing routine MRI in patients with FXS, in most cases any abnormalities are not observed. Therefore, to study the functional organisation of the brain and changes in neuronal activity, we used fMRI.

2. Materials and methods

This study was performed at the Institute International Tomography Centre of the Russian Academy of Sciences (Novosibirsk). A total of 16 male subjects were included in this study, consisting of 8 patients with FXS and 8 healthy volunteers. The study was approved by the local ethical committee. The whole experimental process accorded with the Declaration of Helsinki. The basic procedures and requirements of the experiments were explained to the participants or their parents, and they signed written informed consent forms. MRI exclusion criteria included in a group of healthy volunteers were: metallic implants such as surgical prostheses; history of traumatic brain injury or surgery; history of any chronic, mental, or nervous system disease; availability of permanent pacemakers; claustrophobia.

2.1. FMRI data acquisition

The fMRI was performed on an Achieva (Philips) scanner with a magnetic field strength of 1.5 T by using an 8-channel head SENSE coil. FMRI data were acquired using an echo-planar imaging sequence with the following imaging parameters: matrix = 64×64, 35 slices, voxel size = 4×4×4 mm, repetition time TR = 3500 ms, echo time TE = 50 ms). Reference anatomical images were obtained by T1-TFE (turbo field echo) sequence with the following imaging parameters: matrix = 256×256, 64 slices, voxel size = 1×1×3 mm in three orthogonal projections.

The study design involved registration of the BOLD signal in the resting state. Resting-state fMRI could not be performed in awake patients with FXS owing mental retardation. Therefore, patients with FXS were investigated under sedation with constant control of respiratory and circulatory functions. Propofol® was entered by titration (2.3 mg/kg) and a supporting dose at an infusion rate of 9–15 mg/kg/h. Healthy volunteers were instructed to rest with eyes closed.

2.2. Data analysis

FMRI data were processed with FSL (http://www.fmrib.ox.ac.uk/fsl). Independent component analysis and seed-based correlation analysis were used.

The following pre-statistics processing was applied: slice-timing correction; registration to high-resolution structural and standard space images using FLIRT [2] with further refinement using FNIRT non-linear registration [3]; motion correction using MCFLIRT [2]; and high pass temporal filtering.

Seed-based fMRI data processing was performed using FEAT (FMRI Expert Analysis Tool) Version 6.00, part of FSL. Regions of interest (ROI) time courses were extracted from preprocessed data and then entered into GLM analysis as regressors along with nuisance regressors, which included motion parameters and global, white matter, and CSF average signals. This way it was possible to use GLM to produce whole brain statistical maps of functional connectivity, excluding the nuisance factors mentioned previously.

ROIs were selected from the automatic anatomical labeling (AAL) atlas [4]. Regions related to the cerebellum were not used, resulting in 90 ROIs analysed. Each ROI analyses was performed independently. Resulting statistic images for group comparisons were thresholded using clusters determined by $Z>3.09$ ($p<0.001$) and a corrected cluster significance threshold of $P=0.05$ [5].

Independent component analysis (ICA) was performed using probabilistic independent component analysis [6] as implemented in MELODIC (Multivariate Exploratory Linear Decomposition into Independent Components) Version 3.14, part of FSL. DMN was manually selected out of all identified components. Group comparisons were performed using dual regression procedure [7] to generate subject-specific versions of the spatial maps of the components with further testing for group differences using FSL's randomise permutation-testing tool.

3. Results

3.1. FMRI results Independent Component Analysis

Independent component analysis (ICA) is useful in identifying neural networks, DMN. ICA separates the signal into spatial and temporal components, allowing better removal of noise and the identification of functionally connected submodules within the whole brain.

The DMN subnetwork was successfully identified in both the FXS group and the control group. Three orthogonal projections of DMN identified in the group of patients with FXS are shown in Figure 1. Comparison of results for the two groups showed no statistically significant differences in DMN organisation between patients with FXS and the control group ($p<0.05$).

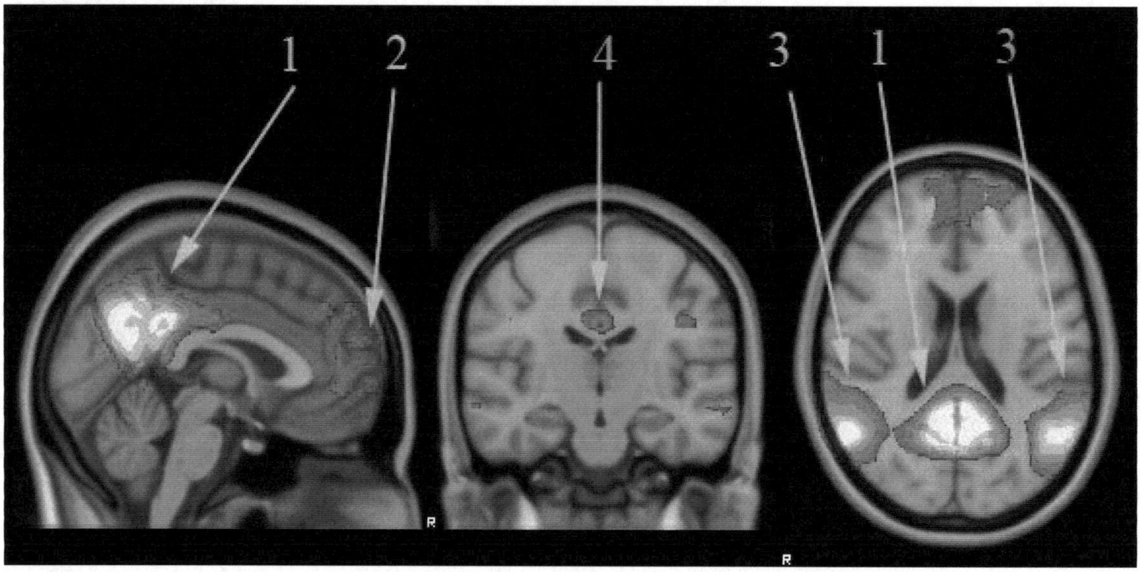

Figure 1. Regions of spontaneous neuronal activity within the default mode network in three projections in patients with FXS. 1 –posterior part cingulate, precuneus, 2 –medial frontal lobes, 3 – parietal lobe, 4 – cingulate gyrus.

3.2. Results of Seed-Based Analysis

Using FSL software, the maps of functional connectivity (FC) for 90 anatomical regions of the brain were obtained for patients of both groups. Statistically significant between-group differences were observed in FC maps for right inferior parietal gyrus and right angular gyrus.

3.2.1. Right inferior parietal gyrus

When comparing the functional connectivity between the groups, it was observed that patients with FXS showed a significantly higher ($p<0.001$) FC between the right inferior parietal and the following brain regions: postcentral, supramarginal gyrus, Brodmann area 2 (which is located in the area of the postcentral gyrus) and Brodmann area 40 (which is located in the area of the supramarginal gyrus).

The resulting map of between-group comparison of the functional connectivity of the healthy volunteers and the FXS group is presented in Figure 2.

3.2.2. Right angular gyrus
When comparing the functional connectivity between the groups, it was observed that patients with FXS showed a significantly higher ($p<0.001$) FC between the right angular gyrus and the following brain regions: cingulate gyrus, limbic lobe, paracentral lobule, medial frontal gyrus, Brodmann area 24 (is part of the anterior cingulate), and Brodmann area 31 (also known as dorsal posterior cingulate area 31). The resulting map of between-group comparison of the functional connectivity of the healthy volunteers and the FXS is presented in Figure 3.

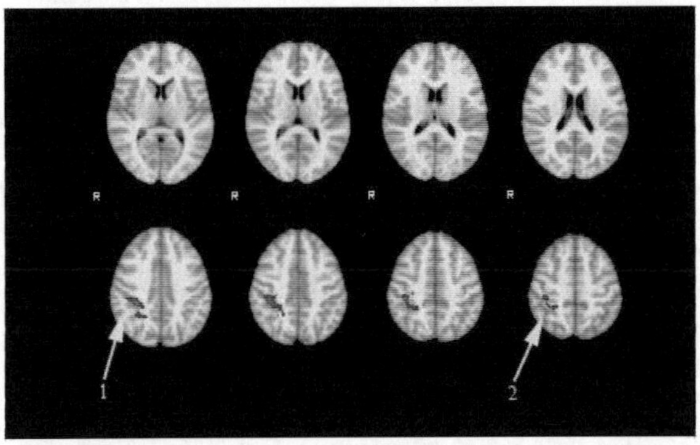

Figure 2. Comparison of functional connectivity in patients with FXS and healthy volunteers. Red colour marks regions in which the spontaneous activity during rest was in higher coherence with that of the region of interest in patients with FXS than in healthy volunteers. As a region of interest - right inferior parietal. 1 – supramarginal gyrus, 2 – postcentral gyrus. p <0.001.

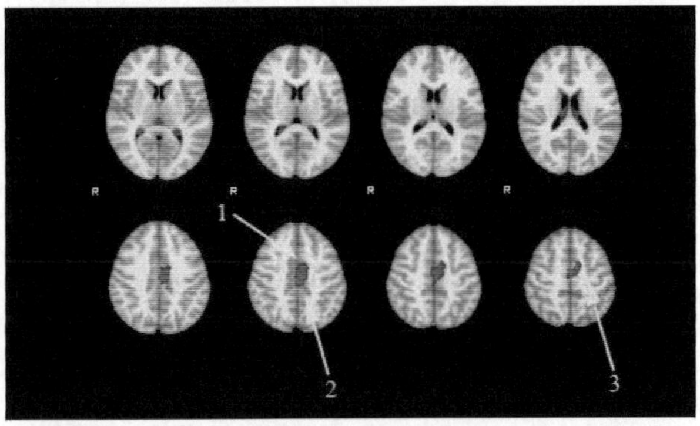

Figure 3. Comparison of functional connectivity in patients with FXS and healthy volunteers. Red colour marks regions in which the spontaneous activity during rest was in higher coherence with that of the region of interest in patients with FXS than in healthy volunteers. As a region of interest - right angular gyrus. 1 –cingulate gyrus, 2 –paracentral lobule, 3 – medial frontal gyrus. p <0.001.

4. Discussion
The high prevalence of FXS, character of inheritance, and development of persistent disabling mental retardation make this disease one of the most important medical and social problems. There have been a number of publications on functional neuroimaging in patients with FXS, but they have all been performed using task-based fMRI paradigms. Also, adult women in the permutation with adult males have been investigated in previous studies [8].

Resting-state fMRI evaluates low-frequency oscillations (0.01-0.1 Hz) in BOLD signal, which is based on the change in hemodynamic parameters that indirectly determines the spontaneous neuronal activity of the different regions that form the neuronal network. When processing resting-state data, temporal similarity and frequency characteristics of oscillations in BOLD signal are estimated for all the voxels of the observer image. It should be noted that the signal of interest for resting-state fMRI analysis is low-frequency oscillations (0.01-0.1 Hz) of time sequences. Currently, there is significant evidence that these low-frequency vibrations are not artefacts at rest (due to respiratory movements and heartbeat) and reflect the important integrative indicator of brain activity–basal activity of neurons. As already stated, DMN is characterised by its activity increase at rest and, conversely, decrease when performing cognitive tasks. This phenomenon represents the occurrence of basal activity of brain neurons in the resting state. It is believed that DMN is involved in cognitive processes and the formation of awareness of one's personality, inner speech, spontaneous thoughts, and dreams. Thus, DMN may affect the formation of consciousness. In addition, in patients with mild cognitive impairment, a reduction in functional relationships within DMN is described compared with healthy volunteers. This fact may have clinical significance if we assume that the functional connectivity between the regions of DMN is a marker of cognitive abilities; the strength of this connection may be of diagnostic value for a variety of clinical conditions and diseases.

Our study using the independent component analysis of resting-state fMRI demonstrates the presence of a DMN in both groups of patients. In our study, DMN brings together the prefrontal, anterior cingulate, posterior cingulate cortex, inferior temporal gyrus, and the superior parietal lobule, which is consistent with previous studies [9]. In this study using group analysis, it was found that the observed DMN did not differ ($p>0.05$) between the two groups of patients, which may indicate the presence of basal neuronal activity in patients with FXS similar to that of control group patients.

When comparing the functional connectivity between the groups, it was observed that patients with FXS showed higher functional connectivity of the right inferior parietal gyrus with postcentral and supramarginal gyri, and Brodmann areas 2 and 40. In addition, it was observed that patients with FXS showed higher functional connectivity between the right angular gyrus and areas located in the cingulate gyrus, limbic lobe, middle frontal gyrus, paracentral lobule, and Brodmann areas 24 and 31. Probable cause of these findings may be a failure in mechanisms of deactivation of certain neural circuits in patients with FXS.

5. Conclusion

New data of functional status of the brain and genetics in patients with FXS were received. Fragile site FRAXA in the patients' blood samples was determined. Presence of the DMN was demonstrated using resting-state fMRI in both the control group and in patients with FXS. Furthermore, it was found that the two groups do not show statistically significant differences ($p>0.05$) in DMN, which may indicate that the basal activity of neurons in patients with FXS is not reduced. Also, an increase in functional connectivity of the right inferior parietal and right angular gyrus ($p<0.001$) in the resting state in patients with FXS was found. Data from this study help shed light not only on the genetic basis, but also on the state of functional connections, in the brain of patients with FXS.

Acknowledgments

The work was financially supported by Russian Science Foundation (the project № 15-15-10001 regarding molecular genetic part, the project №14-35-00020 for works of MRI and the project № 14-45- 00040-П for manuscript preparation and publication) and FASO Russia (project № 0333-2014-0003) regarding theoretical part.

References

[1] McLennan Y, Polussa J, Tassone F and Hagerman R 2011 *Curr Genomics* **12**(3) 216–24
[2] Jenkinson M, Bannister P, Brady M and Smith S 2002 *NeuroImage* **17**(2) 825-841
[3] Andersson J, Jenkinson M and Smith S 2007 Non-linear optimization. FMRIB technical report TR07JA1 (Univ of Oxford FMRIB Centre: Oxford UK)
[4] Worsley K 2002 Statistical analysis of activation images *Functional Magnetic Resonance Imaging: An Introduction to Methods*, ed. P Jezzard, P Matthews and S Smith (York: Oxford Univ. Press) chapter 14 pp 251–270
[5] Tzourio-Mazoyer N, Landeau B, Papathanassiou D, Crivello F, Etard O, Delcroix N, Mazoyer B and Joliot M 2002 *NeuroImage* **15**(1) 273–289
[6] Beckmann C and Smith S 2004 *IEEE Trans Med Imag* **23**(2) 137-152
[7] Filippini N, MacIntosh B, Hough M, Goodwin G, Frisoni G, Smith S, Matthews P, Beckmann C and Mackay C 2009 *PNAS* **106** (17) 7209-14
[8] Tamm L, Menon V, Johnston C, Hessl D and Reiss A 2002 *J Cogn Neurosci* **14**(2) 160–171
[9] Greicius M, Krasnow B, Reiss A and Menon V 2003 *Proc Natl Acad Sci U.S.A* **100** (1) 253–258

IOP Conf. Series: Journal of Physics: Conf. Series **886** (2017) 012008 doi:10.1088/1742-6596/886/1/012008

Dynamics of myelin content decrease in the rat stroke model

A Kisel[1], M Khodanovich[1], D Atochin[1,2], L Mustafina[3] and V Yarnykh[1,4]

[1] Research Institute of Biology and Biophysics, Tomsk State University, Tomsk, Russia
[2] Cardiovascular Research Center and Cardiology Division, Massachusetts General Hospital, Harvard Medical School, Charlestown, Massachusetts, USA
[3] Siberian State Medical University, Tomsk, Russia
[4] Vascular Imaging Laboratory, Department of Radiology, University of Washington, Seattle, WA, USA

E-mail: kisell.alena@gmail.com

Abstract. The majority of studies were usually focused on neuronal death after brain ischemia; however, stroke affects all cell types including oligodendrocytes that form myelin sheath in the CNS. Our study is focused on the changes of myelin content in the ischemic core and neighbor structures in early terms (1, 3 and 10 days) after stroke. Stroke was modeled with middle cerebral artery occlusion (MCAo) in 15 male rats that were divided into three groups by time points after operation. Brain sections were histologically stained with Luxol Fast Blue (LFB) for myelin quantification. The significant demyelination was found in the ischemic core, corpus callosum, anterior commissure, whereas myelin content was increased in caudoputamen, internal capsule and piriform cortex compared with the contralateral hemisphere. The motor cortex showed a significant increase of myelin content on the 1st day and a significant decrease on the 3rd and 10th days after MCAo. These results suggest that stroke influences myelination not only in the ischemic core but also in distant structures.

1. Introduction

Vascular diseases of the brain rank second among the leading causes of death after ischemic heart disease [1]. Studies of ischemic stroke focus primarily on neuronal death and recovery [2-5]. However, this disease affects absolutely all types of brain cells, including oligodendrocytes that form myelin sheath [6-8], a component which is crucial for normal functioning of the nervous system. The purpose of this research was to study the change in the degree of myelination of various structures of white and gray matter after modeling local ischemia of the brain.

2. Materials and methods of the research

The experiment was performed on 15 mature male Sprague-Dawley rats weighing 250-270 g. The ischemic stroke model was created by a reversible middle cerebral artery occlusion (MCAo) [2]. A nylon filament with a silicone tip (Doccol, USA) was introduced into the left internal carotid artery, which thus blocked the bloodstream in the MCA. The ischemic episode began with the introduction of the filament to the MCA and lasted 1 hour, after which the filament was removed. The operation was performed under isoflurane anesthesia (1.5-2% isoflurane in oxygen). During the operation, the body temperature of the animal was maintained at 37±0.5°C. After the operation, the animals were randomly divided into three experimental groups: 1 day, 3 days and 10 days. On the expiration of this period, the animals underwent transcardial perfusion with 4% paraformaldehyde under ether anesthesia. The brain was removed, and cryoprotection was performed in 10% and 20% solutions of sucrose, followed with freezing in liquid nitrogen vapor.

In order to assess the myelination after modeling the ischemic stroke, 10-µm-thick brain sections were made on cryotome HM525 (Termoscientific, Germany), which were then stained with Luxol Fast Blue (LFB) to detect myelin. Photomicrographs of sections were taken with the AxioImager Z2 microscope (Carl Zeiss, Germany) and processed with ImageJ software. For the purposes of the study, several structures of white matter (corpus callosum, cc; commissura anterior, ca; and internal capsula, ic) and gray matter (caudoputamen, cp; cortex dorsalis, cxd; cortex ventralis, cxv; and thalamus, thal) were identified, as well as the area of ischemic lesion (L). In each of those, regions of interest (ROI) were selected, within which the optical density (OD) of staining with LFB (normalized by background) was estimated [9]. The contralateral hemisphere was used as the control, the OD of which was taken as 100%. The statistical processing of data was carried out in STATISTICA 10 software, the significance of differences between the hemispheres was assessed using the χ^2 criterion, and the differences at $p<0.05$ were taken as statistically significant.

3. Results

The results are presented in Figures 1 and 2. As expected, ischemia resulted in a decrease in myelin content in the area of lesion itself ($p<0.001$) at all follow-up times, but the LFB OD in the cp zone adjacent to the area of lesion significantly increased during the 1st day after the stroke modeling.

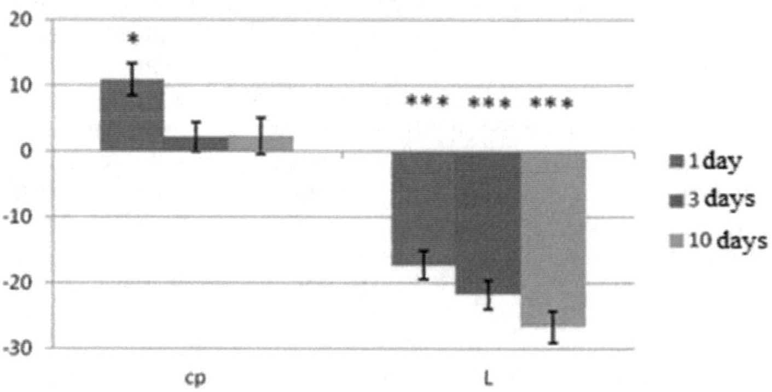

Figure 1. Changes in myelination in the area of ischemic lesion (L) and in caudoputamen of the affected hemisphere (cp) relative to the control hemisphere, expressed as a percentage. The mean ± standard error of the mean. Significant differences between the hemispheres are * – $p<0.05$, *** – $p<0.001$.

Changes in myelination were also observed in the structures not directly affected by ischemia (Figure 2). Regarding the white matter structures, there was a significant decrease in myelin content in cc and ca observed during day 1, which persisted in cc on day 3. On day 10, there were no changes observed in these structures; however, in ic – the area most distant from the ischemic lesion – myelin content increased significantly (Figure 2).

Regarding the gray matter structures, with the exception of thal (the most remote from the zone of ischemic lesion of the structure), there were both an increase in myelination (cxd on day 1 $p<0.05$; and cxv on days 1 and 3 after the surgery $p<0.01$ and $p<0.001$, respectively) and a decrease of this indicator relative to the control (cxd on days 3 and 10 $p<0.01$ and $p<0.05$, respectively) (Figure 2). Apparently, the changes in the degree of myelination of structures over time can be evidence of the redistribution of myelin after ischemic exposure. Besides, in the structures of gray matter, the speed and/or extent of this redistribution can be more pronounced due to the greater water content.

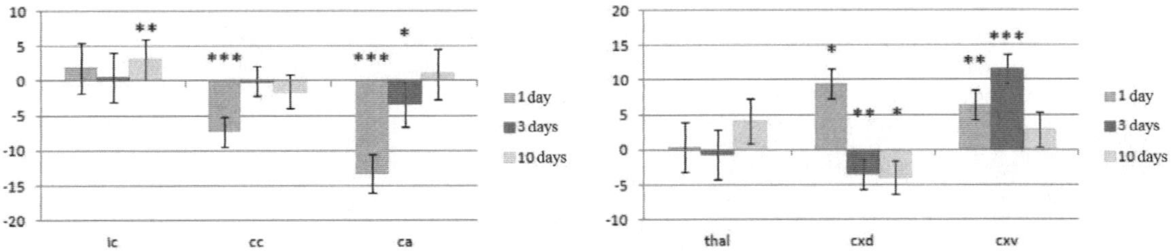

Figure 2. Changes in myelination of the structures of white (left) and gray (right) matter outside the area of ischemic lesion after modeling ischemic stroke relative to the control hemisphere, expressed as a percentage. Notation: internal capsula (ic), corpus callosum (cc), commissura anterior (ca), thalamus (thal), cortex dorsalis (cxd), cortex ventralis (cxv). The mean ± standard error of the mean. Significant differences between the hemispheres are * - $p<0.05$, ** - $p<0.01$, *** - $p<0.001$.

4. Conclusion

The results indicate that in various brain structures that do not even fall under the direct effect of ischemia, there are changes in the degree of myelination, the extent and direction of which depend on various factors, such as the remoteness from the source of ischemia and the initial degree of myelination (white or gray matter). Possible causes of this can be the density of matter in a particular structure, the percentage of water and other less obvious factors.

Acknowledgements

The study was supported by the Russian Science Foundation, project No.14-45-00040-П.

References

[1] The top 10 causes of death: WHO. - http://www.who.int/mediacentre/factsheets/fs310/en/. - 4.02.2017

[2] Khodanovich M, Kisel' A, Chernyshova G, Smol'yakova V, Savchenko R, Plotnikov M 2016 Effect of Fluoxetine on Neurogenesis in Hippocampal Dentate Gyrus after Global Transient Cerebral Ischemia in Rats *Bulletin of Experimental Biology And Medicine* **161(3)** 351-354.

[3] Kisel A, Chernyshova G, Smol'yakova V, Savchenko R, Plotnikov M, Khodanovich M 2016 Modulation of hippocampal neurogenesis in the global brain ischemia model: effects of p-tyrosol and fluoxetine *Annals of Neurology* **80(20)** S265.

[4] Kisel A, Chernysheva G, Smol'yakova V, Savchenko R, Plotnikov M, Khodanovich M 2015 Hippocampal Neurogenesis in the New Model of Global Cerebral Ischemia *AIP Conference Proceedings* **1688** 030004.

[5] Khodanovich M 2015 Reparative neurogenesis after cerebral ischemia: clinical application prospects. *AIP Conference Proceedings* **1688** 030002.

[6] Dewar D, Underhill S, Goldberg M 2003ligodendrocytes and ischemic brain injury. *J Cereb Blood Flow Metab* **23** 263–274.

[7] Jing L, He Q, Zhang J-Z, P. Li A 2013 Temporal profile of astrocytes and changes of oligodendrocyte-based myelin following middle cerebral artery occlusion in diabetic and non-diabetic rats. *Int J Biol Sci* **9** 190-199.

[8] Pantoni L, Garcia J, Gutierrez J. 1996 Cerebral white matter is highly vulnerable to ischemia *Stroke* **27** 1641-1646.

[9] Khodanovich M, Sorokina I, Glazacheva V, Akulov A, Nemirovich-Danchenko N, Romashchenko A, Tolstikova T, Mustafina L and Yarnykh V 2017 Histological validation of

fast macromolecular proton fraction mapping as a quantitative myelin imaging method in the cuprizone demyelination model. *Sci Rep* **7** 46686.

IOP Conf. Series: Journal of Physics: Conf. Series **886** (2017) 012009 doi:10.1088/1742-6596/886/1/012009

The increase in the number of astrocytes in the total cerebral ischemia model in rats

M Kudabayeva[1], A Kisel[1], G Chernysheva[2], V Smol'yakova [2], M Plotnikov[2] and M Khodanovich[1]

[1] Research Institute of Biology and Biophysics, Tomsk State University, Tomsk, Russia
[2] E D Goldberg Institute of Pharmacology and Regenerative Medicine, Tomsk, Russia

E-mail: kmsra08@gmail.com

Abstract. Astrocytes are the most abundant cell class in the CNS. Astrocytic therapies have a huge potential for neuronal repair after stroke. The majority of brain stroke studies address the damage to neurons. Modern studies turn to the usage of morphological and functional changes in astroglial cells after stroke in regenerative medicine. Our study is focused on the changes in the number of astrocytes in the hippocampus (where new glia cells divide) after brain ischemia. Ischemia was modeled by occlusion of tr. brachiocephalicus, a. subclavia sin., a. carotis communis sin. Astrocytes were determined using immunohistochemical labeling with anti GFAP antibody. We found out that the number of astrocytes increased on the 10th and 30th days after stroke in the CA1, CA2 fields, the granular layer of dentate gyrus (GrDG) and hilus. The morphology of astrocytes became reactive in these regions. Therefore, our results revealed long-term reactive astrogliosis in the hippocampus region after total ischemia in rats.

1. Introduction

Stroke is one of the leading causes of death and disability of the population around the world. Elimination of the consequences of this disease is an important task for medicine. The primary attention in the study of ischemic injury mechanisms is directed to neuronal loss and reparative neurogenesis [1-4], while this disease affects absolutely all types of brain cells including astrocytes. Nowadays studies on the development of drugs for the treatment of stroke more and more often become aimed at the support and regulation of astrocytes in this pathology. Astroglia plays an important part in the regulation of post ischemic inflammation [5-8]. This study aimed to investigate the effect of total cerebral ischemia (TCI) on the reactivation of astrocytes.

2. Materials and methods

The experiment was performed on 20 adult male Wistar rats weighing 250-300 g. During the study, the animals were kept in standard vivarium conditions. The model of total ischemia was performed by a surgical occlusion of vessels (left common carotid artery, left subclavian artery and brachycephalic trunk), using ligature [1,3]. The ischemic episode lasted 7 minutes. After the removal of ligatures, reperfusion of vessels ensued. Chloral hydrate anesthesia with artificial pulmonary ventilation was used during the operation, and ether anesthesia – for transcardial perfusion. The tracking of changes was made at several time points. For this purpose, the animals were divided into 4 groups: 10 days after the ischemia – "Control-1" and "Ischemia-1", after 30 days – "Control-2" and "Ischemia-2". The control groups underwent the same manipulations except for the ligature tightening and the creation of an ischemic episode. The fixation of tissues was carried out with a 4% formalin solution. For cryoprotection, the brain was placed in a 10% sucrose solution for 24 hours, followed by a 20% sucrose solution for another 24 hours. After cryoprotection, the tissues were frozen in nitrogen vapors

IOP Conf. Series: Journal of Physics: Conf. Series **886** (2017) 012009 doi:10.1088/1742-6596/886/1/012009

and placed in a freezer for storage at -80°C. Sections of the frozen tissues were prepared using a cryotome (Thermo Scientific HM 525, Germany), at a step of 10 μm. Immunohistochemical analysis of the tissues was carried out using specific antibodies, such as antibodies to GFAP (Rabbit anti GFAP, Alexa Flour, USA) in order to evaluate the reactivation of glia. To perform a fluorescent survey, an antibody marker was used, such as Alexa Flour®488 Donkey anti Rabbit. During the analysis, the number of cells in the hippocampal regions was counted: fields CA1, CA2, layers, granular layer of dentate gyrus (GrDG) and hilus. When carrying out data analysis by means of the STATISTICA 10 software, the t-test for unrelated samples was used.

3. Results

Statistical analysis of the data showed an increase in the number of astrocytes in the "Ischemia-1" group in the CA2 zone, hilus ($p<0.01$), CA1 and GrDG ($p<0.05$), as compared to the "Control-1" group. The reactive proliferation of astrocytes around damaged neurons more than doubled in the fields CA1 and CA2. Anatomically, the fields CA1 and CA2 are characterized with a high density of neurons and a small amount of glia. The migration and proliferation of glia in this zone indicates the presence of inflammation. The least significant changes on day 10 were observed in the GrDG layer, where not only adult astrocytes but also radial glia expressing GFAP underwent the ischemic effect (Fig. 1). These changes indicate the induction of stem cell division.

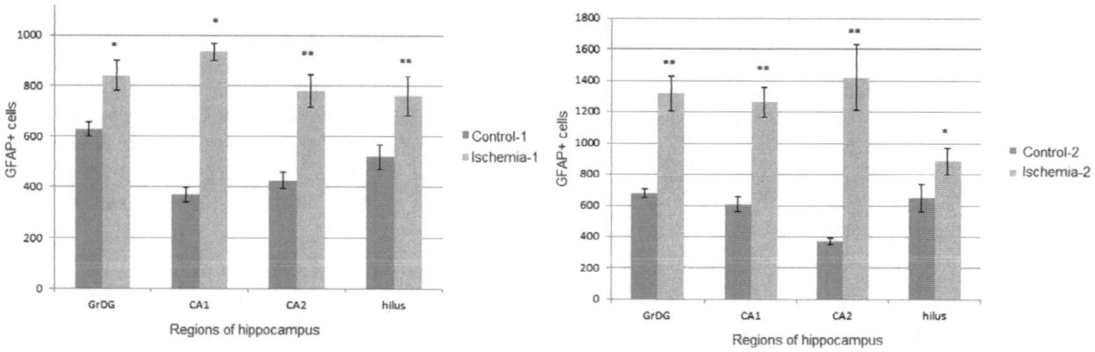

Figure 1. The increase in the number of astrocytes in the layers of the hippocampus after ischemic injury on days 10 and 30. The number of cells per μm^2 is shown. The means for the group with the standard error of the mean are presented. Statistical differences are marked with the following symbols: * – $p<0.05$, ** – $p<0.01$.

Differences between the groups intensified by day 30. There was a significant increase in the number of astrocytes in the layers GrDG, CA1, CA2 ($p<0.01$) and in hilus ($p<0.05$) in the "Ischemia-2" group in comparison with the "Control-2" group. The number of astrocytes in the field CA2 exceeded the control indicators by more than 3 times. The increased differences as compared to the control value in the GrDG layer indicate the presence of a peak in the division of stem cells in the granular layer in the period of about 30 days after ischemia.

In both the ischemic groups, astrocytes retained a morphology characteristic of the state of reactivation, as evidenced by the characteristic 'swelling' of the cells, an increased branching and lengthening of the processes identified by an enhanced expression and GFAP distribution (Figure 2). This morphology persisted at all time points of the study after ischemia.

4. Conclusion

Pyramidal neurons of the CA1 field are most sensitive to ischemic effects [1], and a decrease in blood supply leads to the death of neurons of this layer. The damaged cells of the CA1 field and the signals emitted by them induce the reactivation of astroglia [1,9]. Our previous results showed an increased level of neurogenesis 10 days after total ischemia that however, falls one month after surgery [1-3]. On the contrary, reactive astrogliosis intensifies within a month after ischemia. Based on our data proving the persistence of glial reactivation one month after the ischemic episode, it can be assumed that post-ischemic reactivation of astroglia changes its functions in support of the metabolism of mature neurons in the CA1 layer. The postnatal division of stem cells persists in the GrDG zone of the hippocampus and is a physiologically significant process. Recent studies suggest that the death of neurons is an inducer of proliferation of stem cells which can be differentiated into neuronal progenitors [10]. Our data confirmed that ischemia is a factor stimulating the division of GFAP-positive stem cells.

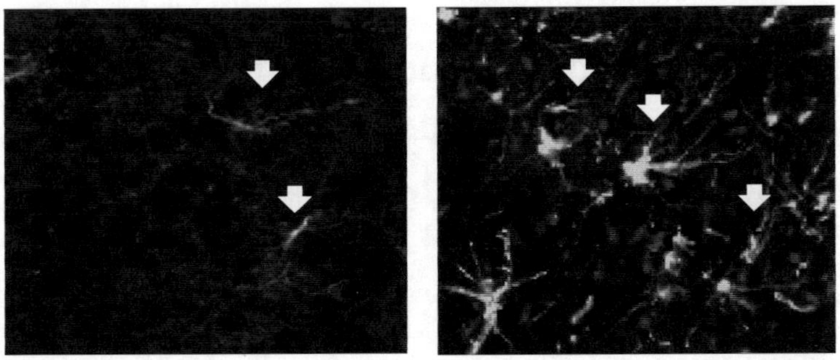

Figure 2. The change in the number of astrocytes in the CA1 field of the hippocampus of rats. The left shows the group "Control-1", and the right – the group "Ischemia-1". The cell nuclei (DAPI) are colored in blue, and the astrocyte processes (GFAP) are colored in green. Image size: 100 μm. Image magnification: × 20.

Acknowledgements
The study was supported by the Russian Science Foundation, project No.14-45-00040-П.

References

[1] Khodanovich M, Kisel' A, Chernyshova G, Smol'yakova V, Savchenko R, Plotnikov M 2016 Effect of Fluoxetine on Neurogenesis in Hippocampal Dentate Gyrus after Global Transient Cerebral Ischemia in Rats *Bulletin of Experimental Biology And Medicine* **161(3)** 351-354.
[2] Kisel A, Chernyshova G, Smol'yakova V, Savchenko R, Plotnikov M, Khodanovich M 2016 Modulation of hippocampal neurogenesis in the global brain ischemia model: effects of p-tyrosol and fluoxetine *Annals of Neurology* **80(20)** S265.
[3] Kisel A, Chernysheva G, Smol'yakova V, Savchenko R, Plotnikov M, Khodanovich M 2015 Hippocampal Neurogenesis in the New Model of Global Cerebral Ischemia *AIP Conference Proceedings* **1688** 030004.
[4] Khodanovich M 2015 Reparative neurogenesis after cerebral ischemia: clinical application prospects. *AIP Conference Proceedings* **1688** 030002.
[5] Pecna M Pecny M 2014 Astrocyte reactivity and reactive astrogliosis: costs and benefits *Physiol. Rev.* **94** 1077 – 1098.
[6] Dossi E, Vasile F and Rouach N 2017 Human astrocytes in diseased brain *Brain Res Bull* pii: S0361-9230(17)30082-5. doi: 10.1016/j.brainresbull.2017.02.001. [Epub ahead of print].
[7] Levine J M, Reynolds R 1999 Activation and proliferation of endogenous oligodendrocyte precursor cells during ethidium bromide-induced demyelination *Exp Neurol* **160** 333–347.
[8] Namihira M and Nakashima K 2013 Mechanisms of astrocytogenesis in the mammalian brain *Current Opinion in Neurobiology* **23** 921–927.
[9] Sims N R and Yew W P 2017 Reactive astrogliosis in stroke: Contributions of astrocytes to recovery of neurological function *Neurochem Int* **107** 88-103.
[10] Lin R, Cai J, Nathan C et al. 2015 Neurogenesis is enhanced by stroke in multiple new stem cell niches along the ventricular system at sites of high BBB permeability. *Neurobiol Dis* **74** 229-239.

Implementation of fast macromolecular proton fraction mapping on 1.5 and 3 Tesla clinical MRI scanners: preliminary experience

V Yarnykh[1,2] and A Korostyshevskaya[3]

[1] Department of Radiology, University of Washington, Seattle, WA, USA
[2] Research Institute of Biology and Biophysics, Tomsk State University, Tomsk, Russian Federation
[3] Institute "International Tomography Center" of the Siberian Branch of the Russian Academy of Sciences, Novosibirsk, Russian Federation

E-mail: yarnykh@uw.edu

Abstract. Macromolecular proton fraction (MPF) is a biophysical parameter describing the amount of macromolecular protons involved into magnetization exchange with water protons in tissues. MPF represents a significant interest as a magnetic resonance imaging (MRI) biomarker of myelin for clinical applications. A recent fast MPF mapping method enabled clinical translation of MPF measurements due to time-efficient acquisition based on the single-point constrained fit algorithm. However, previous MPF mapping applications utilized only 3 Tesla MRI scanners and modified pulse sequences, which are not commonly available. This study aimed to test the feasibility of MPF mapping implementation on a 1.5 Tesla clinical scanner using standard manufacturer's sequences and compare the performance of this method between 1.5 and 3 Tesla scanners. MPF mapping was implemented on 1.5 and 3 Tesla MRI units of one manufacturer with either optimized custom-written or standard product pulse sequences. Whole-brain three-dimensional MPF maps obtained from a single volunteer were compared between field strengths and implementation options. MPF maps demonstrated similar quality at both field strengths. MPF values in segmented brain tissues and specific anatomic regions appeared in close agreement. This experiment demonstrates the feasibility of fast MPF mapping using standard sequences on 1.5 T and 3 T clinical scanners.

1. Introduction

Macromolecular proton fraction (MPF) is a fundamental biophysical parameter defined within the two-pool model of magnetization transfer (MT) and describing the amount of macromolecular protons characterized by quasi-solid molecular dynamics and involved into magnetization exchange with mobile water protons in biological systems [1]. Over past decade, MPF measurements attracted a substantial practical interest due to the discovery of very close correlation between MPF and myelin content in neural tissues [2-6]. For quite a long time, experimental MPF measurements have been unavailable to clinicians, because existing methods (reviewed elsewhere [1]) required extremely time-consuming data acquisition due to the necessity of determining all two-pool model parameters or their subsets from the multi-parameter fit. A recently emerged fast MPF mapping method [7] overcame this difficulty by reducing the number of required images to three by utilizing an optimally constrained model [1], data normalization based on a synthetic (calculated) reference image [7], and an optimal sampling scheme corresponding to an appropriately chosen point in the tissue Z-spectrum [1]. Fast MPF mapping has been successfully tested in clinical studies [8, 9] where it demonstrated a promise as a biomarker of white matter (WM) and grey matter (GM) demyelination in multiple sclerosis [8] and post-concussion syndrome after mild traumatic brain injury [9]. Recently, this method has been histologically validated as a myelin biomarker in the murine cuprizone demyelination model [6] and proposed as a uniform myelin imaging approach for clinical and preclinical studies. The fundamental advantage of MPF is independence of magnetic field strength that enables its application across a range of human and animal imaging platforms [10]. However, the fast MPF mapping method has not been tested on 1.5 T MRI scanners to date, and all published studies utilized MRI equipment with field strengths of 3 T or higher. Implementation of the method for lower fields would considerably extend its clinical availability. Another obstacle to the widespread use of this promising technology is the need in customized pulse sequences to enable optimal saturation of macromolecular protons [1, 7, 8. 9]. Such sequences are difficult or even impossible to implement on each clinical MRI unit. At the same time, the basic acquisition sequence, spoiled gradient echo (GRE) with and without MT preparative module, is commonly available. However, the design of the MT saturation block in manufacturers' product sequences may be substantially suboptimal relative to that needed for accurate MPF measurements according to the original publication [1]. In this preliminary study, we sought to test the feasibility of fast MPF mapping implementation on a 1.5 T clinical scanner using standard manufacturer's sequences and compare the performance of this method between 1.5 T and 3 T scanners of the same manufacturer with both product and customized sequences.

2. Materials and Methods

Experimental data were obtained on the two MRI scanners of the same manufacturer (Philips Medical Systems, Netherlands) including 1.5 T Achieva and 3 T Ingenia units. A 16-channel neurovascular coil and a 32-channel head coil were used on 1.5 and 3 T scanners, respectively. The fast MPF mapping protocol was implemented on the 3 T scanner in two variants. The first option included a custom optimized MT-GRE pulse sequence with the following modifications: replacement of the manufacturer's off-resonance pulse by a more time-efficient one, increased spoiler gradient areas to enable diffusion-based spoiling [11], phase increment for radiofrequency spoiling of 169° [11], standard duration of excitation RF pulse independent of the flip angle, and flexibility in setting MT saturation parameters. The optimized sequence was used for reference measurements and is termed below the "gold standard sequence." The second option included the protocol design based on the unmodified manufacturer's sequences. On the 1.5 T scanner, only the manufacturer's product implementation was used. All protocols included the following sequences required to reconstruct an MPF map using the single-point synthetic reference method [7]:

1) T_1-weighted spoiled GRE imaging: repetition time (TR) = 20 ms, echo time (TE) = 2.3 ms for the 3 T and 6.1 ms for the 1.5 T scanners, flip angle (FA) = 20° and 25° the 3 T and 6.1 ms for the 1.5 T scanners;

2) Proton-density (PD)-weighted spoiled GRE imaging: TR = 20 ms, TE = 2.3 ms for the 3 T and 6.1 ms for the 1.5 T scanners, FA = 3° and 4° the 3 T and 6.1 ms for the 1.5 T scanners;

3) MT-weighted spoiled gradient echo (GRE) imaging: TR = 28 ms for the gold standard sequence at 3 T, 50 ms for the product sequence at 3 T, and 32 ms for 1.5 T scanner, TE = 2.3 ms for the 3 T and 6.1 ms for the 1.5 T scanners, FA = 10° for all protocols.

Off-resonance saturation was achieved by the single-lobe sinc pulse in the gold standard sequence and the three-lobe sinc pulse in the product sequence, both with Gaussian apodization. The following saturation pulse parameters were used in the gold standard sequence: offset frequency 4 kHz, effective FA = 560°, and duration 12 ms. In the product sequence, offset frequency was 1.1 kHz, effective FA was 520° and 1040° for 3 T and 1.5 T, respectively, and duration was 16.5 ms and 15 ms for 3 T and 1.5 T, respectively.

Additionally, for reference measurements, correction of B_0 and B_1 field inhomogeneities was carried out using field maps. For B_0 mapping, the dual-TE GRE phase-difference method [12] was used with $TR/TE_1/TE_2$ = 20/2.3/3.3 ms and FA = 10°. For B_1 mapping, the actual flip-angle imaging (AFI) method [13] was used with $TR_1/TR_2/TE$= 40/160/2.3 ms and FA = 60°.

All images were obtained in 3D mode with single signal acquisition, field-of-view (FOV) = 24x20x24 cm and matrix size 192x134x60 (actual voxel size 1.25x1.5x4 mm) zero-interpolated to 240x240x120. The total scan time was 9 min for the gold standard sequence with additional 6 min for field mapping, 12 min for the product sequence at 3 T, and 10 min for 1.5 T.

Data were acquired from one healthy volunteer. The study was approved by the Ethical Committees at the Tomsk State University and at the International Tomography Center of the Siberian Branch of the Russian Academy of Sciences. Written informed consent was obtained from the participant.

MPF maps were reconstructed according to the single-point synthetic reference image method [7] using custom-written C-language software. MPF maps were registered to the map obtained with the gold standard sequence (rigid registration with six degrees of freedom) and skull stripped using MIPAV software (Center for Information Technology, National Institutes of Health, USA). All MPF maps were then segmented into four tissue classes (white matter (WM), gray matter (GM), mixed WM-GM, and GM mixed with cerebrospinal fluid) using FSL software (Oxford Centre for Functional Magnetic Resonance Imaging of the Brain, University of Oxford, England) as detailed earlier [8]. MPF values were calculated within tissue masks and regions of interest (ROIs) in selected anatomic structures.

3. Results

Table 1 summarizes MPF measurements in segmented brain tissues and anatomic structures obtained using the gold standard sequence and product sequences on 3 T and 1.5 T Philips MRI scanners. Figure 1 illustrates MPF maps obtained in the three described above experiments. Overall, there is very good agreement between MPF values obtained at both field strengths. MPF values measured with the product sequence are also similar to those obtained with the gold standard sequence, though they look slightly lower for certain WM structures. Visually, the MPF maps reconstructed from all datasets appear very similar without any signs of image quality degradation at lower field strength (figure 1).

Table 1. MPF measurements (%) in the human brain on 1.5 T and 3 T MRI scanners.

Tissue, structure	3 T, gold standard sequence	3 T, product sequence	1.5 T, product sequence
Measurements in segmented brain tissues			
WM	11.7	11.3	11.2
GM	5.3	5.2	5.3
Mixed WM-GM	8.0	7.9	8.1
ROI measurements in anatomic structures			
Frontal WM	12.2±0.4	12.2±0.3	11.3±0.4
Occipital WM	12.7±0.4	12.2±0.7	12.0±0.7
Corpus callosum, splenium	13.6±0.4	11.9±0.3	12.6±0.5
Corpus callosum, genu	14.7±0.5	13.5±0.7	13.2±0.8
Pons	11.3±0.4	10.3±0.4	10.5±0.5
Caudate nucleus	5.8±0.4	5.9±0.3	6.1±0.6
Putamen	6.2±0.3	6.5±0.2	6.7±0.2
Insular cortex	5.3±0.6	5.4±0.8	6.0±0.9

4. Discussion

The product implementation of the MT-weighted gradient echo sequence substantially differs from the optimally designed technique [1, 7]. The off-resonance saturation pulse is applied quite close to resonance (1.1 vs. 4 kHz) thus making possible unaccounted saturation of free water protons. The value of the effective flip angle of the saturation pulse is about two-fold greater than that in the optimal setting. Nevertheless, the MPF values obtained with the gold standard and product sequences are essentially similar. Furthermore, the noted aspects of the MT module implementation theoretically might result in overestimation of MPF, whereas the values obtained with the product sequence seem to slightly lower than reference values. The presented example confirms the fact that the single-point MPF mapping method is robust with respect to variations in sequence parameters, because the reconstruction algorithm [1, 7] takes into account actual values of most important parameters including the offset frequency, power, duration, and shape integrals of the saturation pulse, TR, and excitation flip angle.

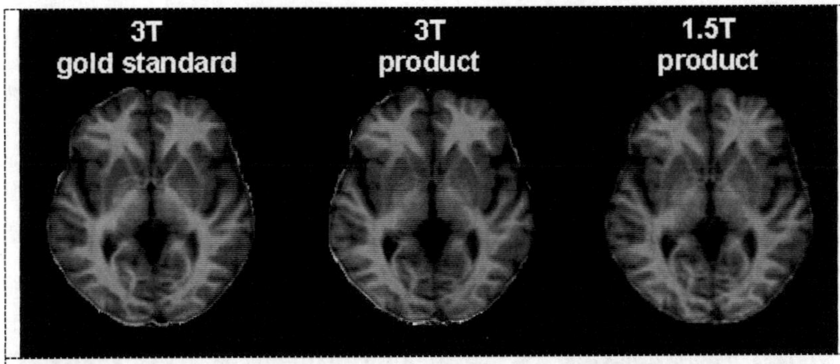

Figure 1. Brain MPF maps of one volunteer obtained using the gold standard sequence on a 3 T scanner and manufacturer's product sequence on 3 T and 1.5 T scanners. All maps are presented with the same grayscale window corresponding to the MPF range 0-20%.

At the same time, there still is a possibility of MPF measurement errors due to hidden from the user sequence implementation details, which are not accounted for in the MPF map reconstruction algorithm. Examples of such factors include variations in the actual excitation bandwidth of the saturation RF pulse, different implementations of RF spoiling, imperfectness of excitation pulses, variable noise level, and some automatic algorithms intended to optimize advanced clinical sequences, such as automatic adjustment of pulse durations dependent on power calibration. Accordingly, instrumental errors may occur in the case of MPF mapping implementation based on the standard manufacturer's sequences without user's control over all relevant parameters. While it is impossible to identify any bias from the single-subject measurements, we speculate that some systematic differences in MPF between the gold standard and product sequences may arise from the suboptimal implementation of radiofrequency spoiling, which is based on the 150° phase increment in the manufacturer's sequence. As it was demonstrated in the earlier publication [11], this spoiling regime causes effective overestimation of T_1 by the variable flip angle method. The error in T_1 may propagate into MPF and result in underestimation of MPF measurements according to the theory detailed elsewhere [14]. Our preliminary experience summarized in this report suggests that such errors are small if not negligible.

5. Conclusion

The described proof-of concept experiment demonstrates the feasibility of fast MPF mapping implementation based on product sequences on 1.5 T and 3 T clinical scanners and high consistency of MPF measurements between field strengths and pulse sequence implementations.

Acknowledgments

This study was supported by the Ministry of Education and Science of the Russian Federation within the State Assignment Project No. 18.2583.2017/4.6. Dr. Korostyshevskaya received salary support from the Federal Agency of Scientific Organizations of the Russian Federation (Project No. 0333-2014-0003).

References

[1] Yarnykh V L 2012 Fast macromolecular proton fraction mapping from a single off-resonance magnetization transfer measurement. *Magn. Reson. Med.***68** 166-178
[2] Ou X, Sun S W, Liang H F, Song S K and Gochberg D F 2009 The MT pool size ratio and the DTI radial diffusivity may reflect the myelination in shiverer and control mice. *NMR Biomed.* **22** 480-487
[3] Underhill H R, Rostomily R C, Mikheev A M, Yuan C and Yarnykh V L 2011 Fast bound pool fraction imaging of the in vivo rat brain: Association with myelin content and validation in the C6 glioma model. *Neuroimage* **54**, 2052-2065
[4] Samsonov A, Alexander A L, Mossahebi P, Wu Y C, Duncan I D and Field A S 2011 Quantitative MR imaging of two-pool magnetization transfer model parameters in myelin mutant shaking pup. *Neuroimage* **62** 1390-1398
[5] Janve V A, Zu Z, Yao S Y, Li K, Zhang F L, Wilson K J, Ou X, Does M D, Subramaniam S and Gochberg D F 2013 The radial diffusivity and magnetization transfer pool size ratio are sensitive markers for demyelination in a rat model of type III multiple sclerosis (MS) lesions. *Neuroimage* **74** 298-305
[6] Khodanovich M Y, Sorokina I V, Glazacheva V Y, Akulov A E, Nemirovich-Danchenko N M, Romashchenko A V, Tolstikova T G, Mustafina L R and Yarnykh V L 2017 Histological validation of fast macromolecular proton fraction mapping as a quantitative myelin imaging method in the cuprizone demyelination model. *Sci. Rep.* **7** 46686
[7] Yarnykh V L 2016 Time-efficient, high-resolution, whole brain three-dimensional macromolecular proton fraction mapping. *Magn. Reson. Med.* **75** 2100-2106
[8] Yarnykh V L, Bowen J D, Samsonov A, Repovic P, Mayadev A, Qian P, Gangadharan B, Keogh B P, Maravilla K R and Henson L K 2015 Fast Whole-Brain Three-dimensional Macromolecular Proton Fraction Mapping in Multiple Sclerosis. *Radiology* **274** 210-220
[9] Petrie E C, Cross D J, Yarnykh V L, Richards T, Martin N M, Pagulayan K, Hoff D, Hart K, Mayer C, Tarabochia M, Raskind M, Minoshima S and Peskind E 2014 Neuroimaging, Behavioral, and Psychological Sequelae of Repetitive Combined Blast/Impact Mild Traumatic Brain Injury in Iraq and Afghanistan War Veterans. *J. Neurotrauma* **31** 425-436
[10] Naumova A V, Akulov A E, Khodanovich M Y and Yarnykh V L 2017 High-resolution three-dimensional macromolecular proton fraction mapping for quantitative neuroanatomical imaging of the rodent brain in ultra-high magnetic fields. *Neuroimage* **147** 985-993
[11] Yarnykh V L 2010 Optimal radiofrequency and gradient spoiling for improved accuracy of T1 and B1 measurements using fast steady-state techniques. *Magn. Reson. Med.* **63** 1610-1626
[12] Skinner T E and Glover G H 1997 An extended two-point Dixon algorithm for calculating separate water, fat, and B0 images. *Magn. Reson. Med.* **37** 628-630
[13] Yarnykh V L 2007 Actual flip-angle imaging in the pulsed steady state: a method for rapid three-dimensional mapping of the transmitted radiofrequency field. *Magn. Reson. Med.* **57** 192-200
[14] Yarnykh V L 2002 Pulsed Z-spectroscopic imaging of cross-relaxation parameters in tissues forhuman MRI: theory and clinical applications. *Magn. Reson. Med.* **47** 929-939

Chronic disorders of consciousness: role of neuroimaging

E Kremneva[1], D Sergeev[1], E Zmeykina[1], L Legostaeva[1] and M Piradov[1]

[1]Research Center of Neurology, Moscow, Russia

E-mail: moomin10j@mail.ru

Abstract. Chronic disorders of consciousness are clinically challenging conditions, and advanced methods of imaging for better understanding of diagnosis and prognosis are needed. Recent functional neuroradiological studies utilizing PET and fMRI demonstrated that besides widespread neuronal loss disruption of interconnection between certain cortical networks after the injury may also play the leading role in the development of behaviourally assessed unresponsiveness. Functional and structural connectivity, evaluated by neuroimaging approaches, may correlate with clinical status and may also play prognostic role. Integration of data from various diagnostic modalities is needed for further progress in this area.

1. Introduction

Severe brain injury may end up in a condition when the patient survived coma regains arousal in the absence (or with very little) awareness of himself and surroundings. These conditions are described as chronic disorders of consciousness [1], with the major sub-entities of vegetative state (VS), now also coined unresponsive wakefulness syndrome (UWS) [2], and minimally conscious state (MCS), in turn, subdivided to MCS plus and MCS minus, depending on the level of awareness spared [3]. The underlying cause of the dissociation between wakefulness and awareness seen in this category of patients is believed to be an impairment of the connectivity between structures within the human brain responsible for arousal (i.e., brainstem and hypothalamus) and the content of consciousness (i.e., fronto-parietal cortical network and its connections to the thalamus) [4].

Clinical assessment of consciousness in patients emerging from coma is limited to bedside evaluation of behavioral response to external stimuli, such as purposeful movements, with the rate of misdiagnosis as high as 37–43% [5]. However, advances in neuroimaging made it possible to study brain structures related to conscious behavior, and, moreover, their intrinsic functional connections and interactions that underlie cognitive processing [6].

2. Structural brain imaging

Conventional structural imaging, such as computed tomography or magnetic resonance imaging (MRI) with standard sequences (T1-TSE, T2-TSE, FLAIR) is used to evaluate the localization and severity of the lesion (traumatic or anoxic; Figure 1), but it fails to indicate a specific brain region(s) that can be undoubtedly related to awareness [7]. More sophisticates methods, such as diffuse tensor imaging (DTI) are utilized to study the integrity of white matter – pathways interconnecting cortical and subcortical structures, thus evaluating *structural connectivity* of the areas relevant to consciousness. DTI techniques are based on the assessment of water diffusion properties and are able to reveal structural damage in normally appearing tissue. They can be applied along with sedation to reduce acquisition artifacts due to the patient's movements, as sedation does not affect the data, which is an important limitation in functional imaging [7].

IOP Conf. Series: Journal of Physics: Conf. Series **886** (2017) 012011 doi:10.1088/1742-6596/886/1/012011

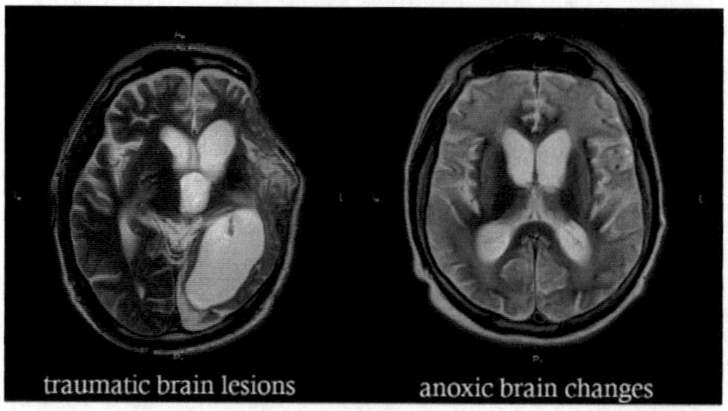

Figure 1. An example of traumatic and anoxic brain lesions leading to chronic disorders of consciousness.

One of the main measurements of the DTI is the fractional anisotropy (FA) that indicates the anisotropy of a diffusion process and describes the integrity of nerve fibers and their direction. It make take values from 0 (isotropic diffusion) to 1 (completely anisotropic diffusion). In the study by Newcombe et al. [8] alterations of FA in VS patients were revealed in supratentorial area (thalamus and corpus callosum) without significant differences in the infratentorial structures (pons and midbrain areas) and were shown to correlate with clinical assessment as measured by Coma Recovery Scale Revised (CRS-R) and functional magnetic resonance imaging (fMRI) activation following an auditory task. DTI showed interconnections between medial prefrontal, medial parietal, and medial temporal cortices within the default network, that is supposed to maintain consciousness [9]. Quantitative DTI techniques may predict clinical outcome after 1 year of traumatic [10] or anoxic [11] injury at the individual-patient level better than structural and clinical assessment (Figure 1).

3. Functional brain imaging

Neurovisualization techniques assessing brain functioning are based on the measurement of the metabolism of brain tissue either directly, as positron emission tomography (PET) with ^{18}Fluorodesoxyglucose (FGD) radioactive tracer does, or via evaluating the blood flow with $^{15}H_2O$-PET or with functional MRI, that quantifies blood-oxygen-level dependent (BOLD) changes.

3.1. PET studies

FDG-PET studies were the first to demonstrate massive decreases in brain metabolism in disorders of consciousness. The reduction of brain function in VS was shown to be 40–50% of normal values [12]. Later it was shown that some cortical areas are more important in respect with consciousness than others, as patients in VS demonstrated decreased metabolism in a widespread network encompassing frontoparietal areas [13]. Clinical recovery from the UWS did not coincide with the recovery of global metabolic levels; rather, it was paralleled with *functional connectivity* restoration in these areas (lateral prefrontal and posterior parietal regions, midline anterior cingulate/mesiofrontal and posterior cingulate/precuneal associative cortices) and between these regions and the thalamus [14]. These findings lead to the description of awareness network, that may be subdivided into the intrinsic, or default mode network (DMN), and extrinsic, or executive control network (ECN). The latter includes the lateral fronto-parietal brain regions and is related to sensory awareness or awareness of the environment, while the former encompasses the medial prefrontal cortex and the precuneus and bilateral posterior parietal cortices and is responsible to internal awareness [15]. The discovery of these networks gave rise to the understanding that the disorders of consciousness in certain patients represent the disconnection syndromes rather than the consequences of extensive neuronal loss. In VS/UWS patients PET-measured activation in response to noxious stimulus was limited to the primary sensory areas [16], that were disconnected from associative fronto-parietal cortical networks, which are thought to be essential for conscious perception. At the same time, MCS patients demonstrated activation of higher order associative areas, that indicates partial preservation of this networks integrity [17], while midline regions may remain highly dysfunctional [18]. PET studies also revealed activation in response to the pain in MCS patients similar to that observed in healthy controls, suggesting a possible perception of pain in this patient category. Metabolism in the awareness network was shown to correlate to the CRS-R score [19].This data suggests that patients in MCS may retain altered self-awareness besides their behavioral response that may be very limited, e.g., due to the motor deficit or aphasia. However, FDG-PET cannot yet disentangle between VS/UWS and MCS at the single-subject level.

3.2. Functional MRI studies

Higher availability of fMRI led to gradual replacement of the PET activation studies in the recent years. Active fMRI paradigms confirmed findings obtained with PET and revealed high level cortical activation in response to visual, auditory and somatosensory stimuli in MCS patients, while only low level cortical activation, limited to the primary sensory areas, was detected in VS/UWS [22]. The VS/UWS patients who showed high level cortical activation later often demonstrated signs of recovery [23]. Furthermore, fMRI studies demonstrated the ability to willfully modulate brain activity in the minority of patients who do not show any clinical signs of awareness. Monti et al. [20] reported clinically VS patient who were able to correctly respond to the questions by imaging different situations that was captured by fMRI. This and several following studies [21] confirm the shortcomings of solely clinical assessment of this category of patients. However, despite certain diagnostic and prognostic potential, fMRI still cannot be viewed as a "golden standard" for covert consciousness detection in behaviorally unresponsive patients, as the absence of cortical activation to certain stimuli does not necessarily coincide with absence of awareness.

Another fMRI approach is used to investigate the spontaneous temporal coherence in BOLD fluctuations related to the amount of synchronized neural activity (i.e., functional connectivity) in the absence of any external stimuli (resting-state fMRI). It allows to detect an activity of several resting state networks, such as the DMN, bilateral fronto-parietal or executive control networks, salience, sensorimotor, auditory, visual systems, and the cerebellar network [6]. The connectivity of DMN was shown to correlate with the level of consciousness according to literature [24] and our own expirience (Figure 2).

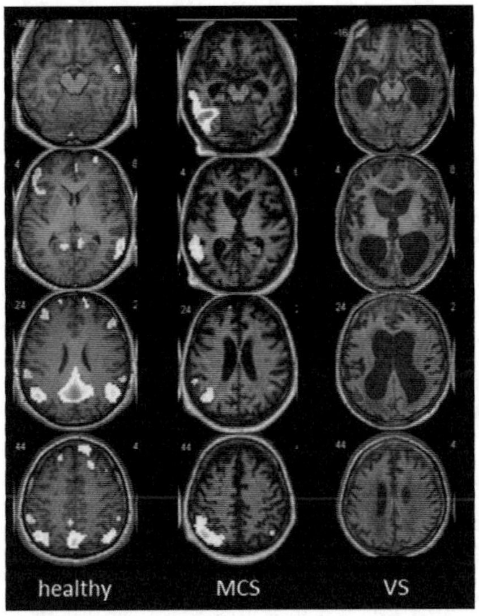

Figure 2. Defaut mode network (resting-state fMRI) in healthy subject (all components showed), partially preserved DMN in minimal conscious state and almost absence of DMN in vegetative state

However, at the single-patient level, fMRI still fails to reliably distinguish VS/UWS from MCS. Recently, resting-state fMRI was used to investigate functional connectivity between cortical regions and brainstem with the view of finding sites crucial for coma development [25]. A small region in the rostral dorsolateral pontine tegmentum functionally connected to the ventral anterior insula and pregenual anterior cingulate cortex was significantly associated with coma-causing lesions that may be considered as a new functional network relevant to maintaining consciousness.

4. Conclusion

Recent progress in neuroimaging techniques played key role in the understanding of the disorders of consciousness. Structural and functional imaging is the source of insights into intrinsic brain networks interaction crucial for conscious behavior. The next step of development will be the integration of data acquired with different modalities, including neuroimaging and neurophysiological approaches (e.g., TMS-EEG, with the assessment of *effective connectivity*), to develop the single-patient prognostic factors and to clear the paths to treatment and recovery.

Acknowledgments
E. Kremneva, E. Zmeykina, L. Legostaeva, M. Piradov are supported by Russian Science Foundation grant No 16-15-00274.

References

[1] Posner J B, Saper C B, Schiff N D and Plum F 2007 *Plum and Posner's diagnosis of stupor and coma* (Oxford University Press)
[2] Laureys S et al 2010 European Task Force on Disorders of Consciousness. Unresponsive wakefulness syndrome: a new name for the vegetative state or apallic syndrome. *BMC Med.* **8** 68
[3] Giacino J T et al 2002 The minimally conscious state: definition and diagnostic criteria *Neurology* **58(3)** 349–353
[4] Di Perri C, Stender J, Laureys S and Gosseries O 2014 Functional neuroanatomy of disorders of consciousness *Epilepsy Behav.* **30** 28-32
[5] Schnakers C, Vanhaudenhuyse A, Giacino J, Ventura M, Boly M, Moonen G and Laureys S 2009 Diagnostic accuracy of the vegetative and minimally conscious state: clinical consensus versus standardized neurobehavioral assessment. *BMC Neurol* **9(1)**
[6] Di Perri C, Thibaut A, Heine L, Soddu A, Demertzi A and Laureys S 2014 Measuring consciousness in coma and related states *World Journal of Radiology* **6(8)** 589-597
[7] Tshibanda L, Vanhaudenhuyse A, Boly M, Soddu A, Bruno MA, Moonen G, Laureys S and Noirhomme Q 2010 Neuroimaging after coma. *Neuroradiology* **52(1)** 15-24
[8] Newcombe V F, Williams G B, Scoffings D, Cross J, Carpenter T A, Pickard J D and Menon D K 2010 Aetiological differences in neuroanatomy of the vegetative state: insights from diffusion tensor imaging and functional implications *J Neurol Neurosurg Psychiatry* **81(5)** 552-61
[9] Greicius M D, Supekar K, Menon V and Dougherty R F 2009 Resting-state functional connectivity reflects structural connectivity in the default mode network *Cereb Cortex* **19(1)** 72-8
[10] Galanaud D et al 2012 Neuro Imaging for Coma Emergence and Recovery Consortium. Assessment of white matter injury and outcome in severe brain trauma: a prospective multicenter cohort *Anesthesiology* **117(6)** 1300-10
[11] Luyt C E et al 2012 Neuro Imaging for Coma Emergence and Recovery Consortium. Diffusion tensor imaging to predict long-term outcome after cardiac arrest: a bicentric pilot study *Anesthesiology* **117(6)** 1311-21
[12] Laureys S 2005 The neural correlate of (un)awareness: lessons from the vegetative state *Trends Cogn Sci.* **9(12)** 556-9
[13] Nakayama N, Okumura A, Shinoda J, Nakashima T and Iwama T 2006 Relationship between

regional cerebral metabolism and consciousness disturbance in traumatic diffuse brain injury without large focal lesions: an FDG-PET study with statistical parametric mapping analysis *J Neurol Neurosurg Psychiatry* **77(7)** 856-62
[14] Laureys S, Lemaire C, Maquet P, Phillips C and Franck G 1999 Cerebral metabolism during vegetative state and after recovery to consciousness *J Neurol Neurosurg Psychiatry* **67(1)** 121
[15] Vanhaudenhuyse A et al 2011 Two distinct neuronal networks mediate the awareness of environment and of self. *J Cogn Neurosci.* **23(3)** 570-8
[16] Boly M et al 2008 Perception of pain in the minimally conscious state with PET activation: an observational study *Lancet Neurol.* **7(11)** 1013-20
[17] Laureys S, Perrin F, Faymonville ME, Schnakers C, Boly M, Bartsch V, Majerus S, Moonen G and Maquet P 2004 Cerebral processing in the minimally conscious state *Neurology* **63(5)** 916-8
[18] Thibaut A et al 2012 Metabolic activity in external and internal awareness networks in severely brain-damaged patients *J Rehabil Med.* **44(6)** 487-94
[19] Giacino J T, Fins J J, Laureys S and Schiff N D 2014 Disorders of consciousness after acquired brain injury: the state of the science *Nat Rev Neurol.* **10(2)** 99-114
[20] Monti M M, Vanhaudenhuyse A, Coleman M R, Boly M, Pickard J D, Tshibanda L, Owen A M and Laureys S 2010 Willful modulation of brain activity in disorders of consciousness. *N Engl J Med* **362(7)** 579-89
[21] Naci L and Owen A M 2013 Making every word count for nonresponsive patients. *JAMA Neurol.* **70(10)** 1235-41
[22] Di H, Boly M, Weng X, Ledoux D and Laureys S 2008 Neuroimaging activation studies in the vegetative state: predictors of recovery? *Clin Med (Lond)* **8(5)** 502-7
[23] Di H B, Yu S M, Weng X C, Laureys S, Yu D, Li J Q, Qin P M, Zhu Y H, Zhang S Z and Chen Y Z 2007 Cerebral response to patient's own name in the vegetative and minimally conscious states *Neurology* **68(12)** 895-9
[24] Vanhaudenhuyse A et al 2010 Default network connectivity reflects the level of consciousness in non-communicative brain-damaged patients *Brain* **133(Pt 1)** 161-71
[25] Fischer D B et al 2016 A human brain network derived from coma-causing brainstem lesions *Neurology*. Nov 4. [Epub ahead of print]

IOP Conf. Series: Journal of Physics: Conf. Series **886** (2017) 012012 doi:10.1088/1742-6596/886/1/012012

MRI evaluation of frequent complications after intra-arterial transplantation of mesenchymal stem cells in rats

D Namestnikova[1], I Gubskiy[1,2], A Gabashvili[3], K Sukhinich[4], P Melnikov[5], D Vishnevskiy[1], A Soloveva[1], E Vitushev[1], V Chekhonin[1,5], L Gubsky[1] and K Yarygin[3]

[1] Pirogov Russian National Research Medical University, Moscow, Russia
[2] Research Institute of Biology and Biophysics, Tomsk State University, Tomsk, Russia
[3] Institute of Biomedical Chemistry, Moscow, Russia
[4] Koltzov Institute of Developmental Biology, Moscow, Russia
[5] Serbsky Federal Medical Research Centre of Psychiatry and Narcology, Moscow, Russia

E-mail: dadnam89@gmail.com

Abstract. Intra-arterial transplantation of mesenchymal stem cells (MSCs) is an effective delivery route for treatment of ischemic brain injury. Despite significant therapeutic effects and targeted cells delivery to the brain infraction, serious adverse events such as cerebral embolism have been reported and may restrict potential clinical applications of this method. In current study, we evaluate potential complications of intra-arterial MSCs administration and determine the optimum parameters for cell transplantation. We injected SPIO-labeled human MSCs via internal carotid artery with different infusion parameters and cell dose in intact rats and in rats with the middle cerebral occlusion stroke model. Cerebrovascular complications and labeled cells were visualized in vivo using MRI. We have shown that the incidence of cerebral embolic events depends on such parameters as cell dose, infusion rate and maintenance of blood flow in the internal carotid artery (ICA). Optimal parameters were considered to be 5×10^5 hMSC in 1 ml of PBS by syringe pump with velocity 100 μ/min and maintenance of blood flow in the ICA. Obtained data should be considered before planning experiments in rats and, potentially, can help in planning clinical trials in stroke patients.

1. Introduction

Ischemic stroke is one of the leading causes of death and disability worldwide. The only currently available and effective treatment is reperfusion methods, that are restricted by "narrow" therapeutic window and have many contraindications [1]. Transplantation of mesenchymal stem cells (MSCs) has shown promising therapeutic results in different animal models of stroke and in humans according to the results of clinical trials [2]. MSCs have several practical advantages for clinical application, including proven safety, lower risks of malignancies and availability. Therefore, transplantation of MSCs can be a promising treatment strategy to reduce cerebral infarction and neurological deficit [3]. However, the cellular and molecular mechanisms of MSCs' beneficial action, as well as the optimum therapeutic time window, route of administration and cell dose still have to be determined.

Intracerebral, intravenous and intra-arterial routes of MSCs transplantation are all effective in ischemic stroke models [4]. Specifically, intra-arterial administration provides targeted delivery of cells to the brain lesion bypassing the peripheral organs and ensuring better functional recovery after stroke [5]. Unfortunately, intra-arterial administration can trigger serious adverse events restricting potential clinical use, such as cerebral embolism [6, 7]. It has been argued that parameters of cell administration such as cell dose, infusion volume, velocity and injection technique are related to the severity and frequency of complications after intra-arterial MSCs delivery [7, 8].

In this study, we evaluate potential complications of intra-arterial MSCs administration and determine the optimum parameters for cell transplantation. We injected SPIO-labeled (superparamagnetic iron oxide microparticles) human MSCs via internal carotid artery with different infusion parameters and cell dose in intact rats and in rats with the middle cerebral occlusion stroke model. Cerebrovascular complications and labeled cells were visualized in vivo using MRI. Obtained data provided recommendation for safety intra-arterial MSCs transplantation and should be considered before planning experiments in rats and, potentially, in stroke patients in clinical trials.

2. Materials and Methods

2.1 Cell culture and labelling

Mesenchymal stem cells (MSCs) were isolated from normal human placenta (gestation age 38-40 weeks). Human placental samples were washed with Hanks solution (PanEko) and after gentle mechanical stimulation incubated with 0.1% solution of type I collagenase (Gibco) for 30 min at 37° C. The suspension was centrifuged (300 × g, 10 min) resuspended in Hanks Solution (Gibco, Invitrogen), and processed for mononuclear fraction separation. The cells were cultured in DMEM-F12, supplemented with 2 mM glutamine, 100 U/ml penicillin, 0.1 mg/ml streptomycin (Gibco) and 10% FBS at 37°C, in a humidified atmosphere containing 5% CO2 in T75 culture flask. Cells were allowed to adhere for 3 days and non-adherent cells were removed by replacing the media. Upon reaching 80% to 90% confluence, adherent MSCs were harvested via trypsinization and subcultured at 1:3 ratio in T75 flask. MSCs were used for labeling and transplantation after second passage.

Human mesenchymal stem cells (hMSCs) were labeled with SPIO-containing microparticles (MC03F, Bang Laboratories, Inc., d=0.90 μ) carrying Dragon Green fluorescent dye (λex=480 nm, λem=520 nm) according to manufacturer's instructions. A suspension of microparticles was added to hMSCs culture when it attained 80-90% confluence (2,5 μl stock suspension per 1 ml culture medium) and incubated for 12h in a CO2 incubator. After incubation, hMSCs were washed twice with PBS to remove the particles that had not been taken up and then harvested by trypsinization.

Additionally, in part of animals to confirm the localization of SPIO microparticles we transplanted hMSCs co-labeled with lipophilic membrane cell tracking red fluorescent dye PKH26 (Sigma-Aldrich) according to the manufacturer's protocol.

2.2 Animals

All manipulations with experimental animals were carried out in accordance with the European Communities Council Directive of 24 November 1986 regarding the protection of animals used for experimental and other scientific purposes (86/609/EEC) and were approved by the local Ethical Committee of the Pirogov Russian National Research Medical University. Adult male Wistar rats (n=46) weighing 250–300 g were used for the experiment. Animals were randomly divided into the following experimental groups depending on the parameters of intra-arterial hMSCs administration: group 1 (n=5) – intact rats with transplantation of $5x10^5$ hMSC in 1 ml of PBS by manual syringe injection with the velocity 300 μ/min and no blood flow in the internal carotid artery (ICA); group 2 (n=6) – intact rats with transplantation of $5x10^5$ hMSC in 1 ml of PBS by syringe pump with velocity 100 μ/min and no blood flow in the ICA; group 3 (n=5) – intact rats with transplantation of $5x10^5$

hMSC in 1 ml of PBS by syringe pump with velocity 100 μ/min and maintenance of blood flow in the ICA; group 4 (n=3) – intact rats with transplantation of $1x10^6$ hMSC in 1 ml of PBS by syringe pump with velocity 100 μ/min and maintenance of blood flow in the ICA; group 5 (n=27) – rats with transplantation of hMSCs 24h after ischemic stroke modelling with previously determined optimal parameters of cell administration (1x105 hMSC in 1 ml of PBS by syringe pump with velocity 100 μ/min and maintenance of blood flow in the ICA).

2.3 Transient middle cerebral artery occlusion

Transient middle cerebral artery occlusion (MCAO) was performed in animals of group 5 according to the intraluminal suture method [9–11]. For the surgery rats were anesthetized with 3% isoflurane and maintained at artificial ventilation with the mixture of 2-2.5% isoflurane and 97.5-98% atmospheric air supplied by E-Z-7000 animal anaesthesia system (E-Z Anaesthesia® Systems), temperature was maintained at 37°C with a heating pad during anaesthesia. Midline incision of the skin and dissection of superficial fascia and muscles were performed to expose the right common carotid artery (CCA), the external carotid artery (ECA) and internal carotid artery (ICA). The vagus nerve was anaesthetized additionally with 0.2 ml of 0.5% bupivacaine and sharply separated from the CCA and ICA. After ligation of the pterygopalatine artery (PPA), cauterization of superior thyroid artery (STA) and occipital artery (OA), and after placing the microsurgical clip on CCA, the ECA was then cut with microscissors and silicon rubber-coated 4-0 monofilament (Doccol corporation, diameter 0.19 mm, length 30 mm; diameter with coating 0.37+/- 0.02 mm; coating length 3-4 mm) was inserted into the stump of the ECA down toward the lumen of ICA for 18-20 mm to the middle cerebral artery (MCA) till mild resistance was felt. Since that time the occlusion of MCA started. The silk suture around the ECA stump was tightened to prevent bleeding, the microclip from the CCA was removed and the incision was closed. After 90 min the rat was re-anesthetized, afterwards the incision was reopened, filament was withdrawn, the stump of the ECA was ligated and the operating wound was closed.

2.4 Intra-arterial hMSCs transplantation

Cell transplantation was performed under isoflurane anaesthesia as described above. The right common carotid artery (CCA), external carotid artery (ECA), and internal carotid artery (ICA) were exposed (or re-exposed 24 h after MCAO in case of group 6). Microsurgical clips were placed on CCA and ICA, while the PPA was ligated (not ligated in part of the animals). The ECA was then cut and a polyurethane microcatheter (Doccol corporation, outer diameter 0.635 mm, inner diameter 0.305 mm) filled with saline was inserted into the stump of the ECA and advanced into the extracranial branch of ICA. The microsurgical clip from CCA was removed to maintain the blood flow in the ICA (for animal group 3-5) of left in CCA till the end of cells infusion (for animal groups 1-2). The catheter was then connected to a 1 ml syringe, for groups 2-6 syringe was placed in the microinjector, and SPIO-labeled hMSCs suspended in 1 ml of PBS were delivered into the ICA over different time intervals. After cell transplantation the stump was electrocoagulated, and incision was closed.

2.5 Magnetic resonance imaging

Before and immediately after hMSCs transplantation, animals of all experimental groups underwent brain MRI examination with the 7T ClinScan system (Bruker BioSpin, USA) with phased array four elements rat brain coil to evaluate cerebrovascular complications. During the procedure rats were maintained at inhalation anaesthesia as described in the previous section. The following MRI protocols were used:

- T2 weighted imaging (T2WI; Turbo Spin Echo pulse sequence with restore magnetization pulse; TR/TE = 4000/46 ms; spectral fat saturation; FOV = 37 x 29.6 mm; slice thickness = 0.5 mm; matrix size = 320 x 256);
- Susceptibility Weighted Imaging (SWI; 3D Gradient Echo with RF spoiling and flow compensation; TR/TE=50/19.1 ms; flip angle = 15; spectral fat saturation; FOV = 30 x 20.6 mm; slice thickness = 0.5 mm; matrix size = 256 x 176).

- Diffusion weighted image with calculation of apparent diffusion coefficient maps (DWI with ADC; echo planar pulse sequence; TR/TE=5000/22 ms; 14 b factors from 0 to 2000 s/mm^2; spectral fat saturation; FOV = 32 x 20 mm; slice thickness = 1 mm; matrix size = 80 x 52).

2.6 Histology
The animals were sacrificed with a lethal dose of Chloral hydrate immediately after MRI and transcardially perfused with 100 ml phosphate buffered saline (PBS, pH 7.4, 0.1 M) followed by 100 ml of ice-cold 4% paraformaldehyde in PBS. Series of coronal 40 μm-thick sections were obtained with a HM 650 v vibratome (Microm GmbH, Berlin, Germany). Staining of cell nuclei was performed by incubating slide-mounted tissue sections for 10 min with 2 μg/mL DAPI solution (Sigma). Fluorescence confocal micrographs were captured with the Nikon A1R MP+ laser scanning confocal microscope equipped with a 405, 488, 561, 638 lasers and Plan Apo20x/0.75 Dic N, Apo IR 60x/1.27 WI and Apo TIRF 60x/1.49 oil Dic objective lenses. To visualize the iron microspheres, Perls' Prussian blue staining was performed and bright-field images were acquired using the Keyence BZ-9000E microscope.

3. Results and Discussion
We found that frequent complications after intra-arterial administration of hMSCs included extracranial distribution of transplanted cells and cellular embolism of the intracranial arteries, followed by the development of cerebral infarctions.

3.1 Extracranial MSCs distribution
Internal carotid artery has a large extracranial branch, the pterygopalatine artery (12 in Figure 1). Consequently, part of cells injected via ICA can be transferred to the PPA blood supply area.

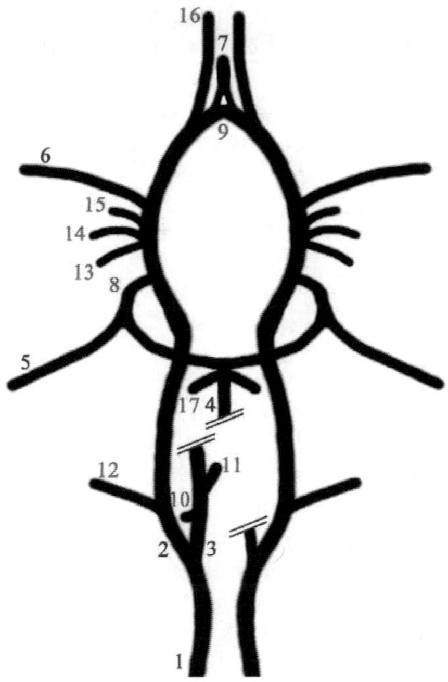

Figure 1. Arterial system of rat brain. Numbers on figure indicate:
1. Common carotid artery
2. Internal carotid artery
3. External carotid artery
4. Basilar artery
5. Posterior cerebral artery
6. Middle cerebral artery
7. Anterior cerebral artery
8. Posterior communicating artery
9. Anterior communicating artery
10. Occipital artery
11. Superior thyroid artery
12. Pterygopalatine artery
13. Hypothalamic artery
14. Ventral thalamic artery
15. Anterior choroidal artery
16. Olfactory artery
17. Superior cerebellar artery

Figure 2 demonstrates the distribution of SPIO-labeled hMSCs not only in the right brain hemisphere, but also in the masticatory muscles. To prevent extracranial "cell leakage", the PPA should be ligated prior to cell infusion.

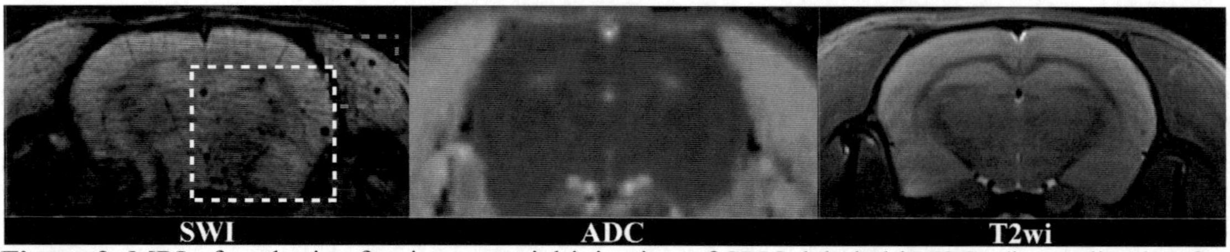

Figure 2. MRI of rat brain after intra-arterial injection of SPIO-labeled hMSCs (immediately after injection). The white dotted box delineates the area of hMSCs distributions in the right brain hemisphere, while the red dotted box shows the area of extracranial cell scattering in the masticatory muscles (zone of PPA blood supply). SPIO-labeled hMSCs could be visualized as hypointense spots with SWI. On ADC and T2wi no embolic complications are visualized.

3.2 Cerebral embolism

We investigated the relationships between cell dose, infusion velocity, injection technique and cerebral embolic events after intra-arterial transplantation of hMSCs. In MRI, embolic stroke areas are visualized as zones of low apparent diffusion coefficient (ADC) on ADC maps in acute stage.

3.2.1 Intact rats. The optimum parameters of transplantation were determined in the experiments with intact rats. Animals were divided into 4 groups, each representing sequential change of the infusion parameters, as described in the Materials and Methods section.

The first studied parameter was the rate of MSCs infusion. The cell concentration of $5x10^5$ hMSCs in 1 ml of PBS chosen for this experiment has been previously evaluated as relatively safe and sufficient to provide therapeutic effect [8]. In group 1 hMSCs were injected via ICA using a manual syringe at the rate of about 300 μ/min, and in group 2 – by a syringe pump at the rate of 100 μ/min. In both groups no blood flow in ICA around the catheter was maintained. As can be seen in Figure 3, in the first two groups severe cerebral embolism followed by ischemia and the development of cerebral infarction was observed (in group 1 in 4 out of 5 rats, and in group 2 in 5 out of 6 rats). In group 3 with the slow infusion rate of 300 μ/min and the opening of blood flow around ICA no embolic complications at all were observed. Probably, the maintenance of blood flow in ICA prevents the hMSCs aggregation and improves their distribution.

SPIO-labeled cells are visualized in the brain tissue as single hypointense spots with SWI, which was confirmed by histological study (displayed in Figure 4 (a) and (b)).

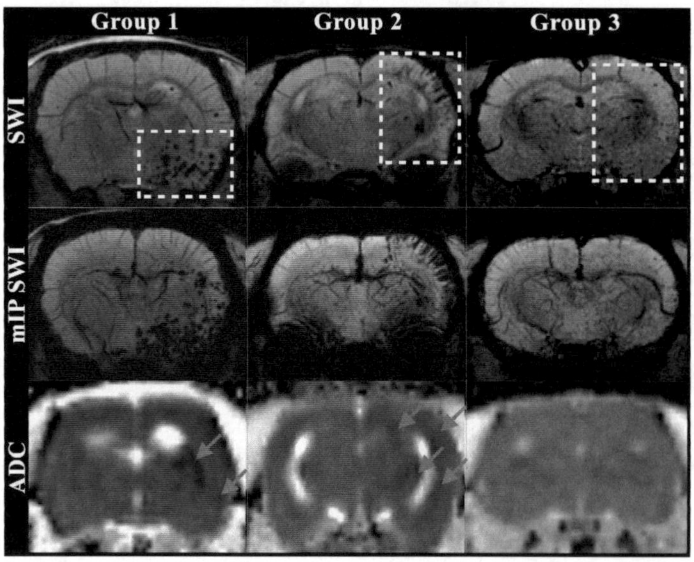

Figure 3. MRI of rat brains immediately after intra-arterial injection of SPIO-labeled hMSCs in groups 1-3. Detailed description of each group is given in the "Materials and Methods" section. White dotted box indicates areas of hMSCs distribution in the right brain hemisphere, and red arrow indicates infarction zone due to cell embolism. SPIO-labeled hMSCs visualized as hypointense spots on SWI.

(a)

(b)

(c)

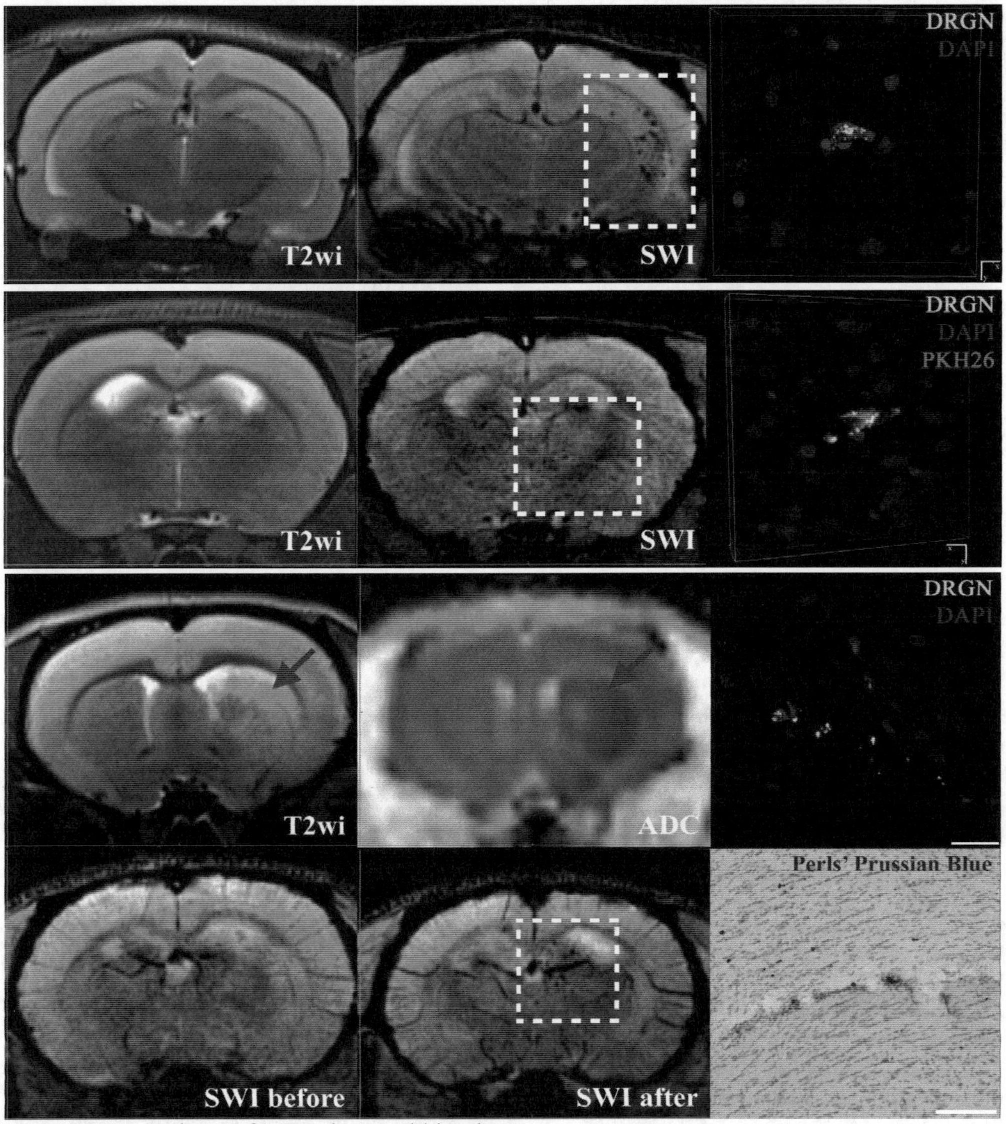

Figure 4. Comparison of MRI data and histology.
(a) – brain images taken after intra-arterial injection of SPIO-labeled hMSCs in a group 3 rat without cellular embolism. Confocal microscopy image shows fluorescently labeled SPIOs (Dragon green - DRGN) and nuclear counterstain with DAPI (blue).
(b) – another example of brain images of a rat from group 3 without cellular embolism, hMSCs have additional label: membrane dye PKH26 is red and SPIO microparticles in cytoplasm are green.
(c) – example of the images of brain taken from a group 5 rat with transplantation of hMSCs 24h after MCAO and optimal parameters of cell administration). Confocal microscopy images showing fluorescent labeled SPIOs (Dragon green – DRGN) and nuclear counterstain with DAPI (blue). Brightfield microscopy (Perl's Prussian blue staining) images showing SPIO label in brain tissue.
White dotted boxes indicate areas of cell distribution in the right brain hemisphere. SPIO-labeled hMSCs visualized as hypointense spots with SWI. Blue arrow indicates infarction zones formed after MCAO. The scale bars represent 20 μm for confocal microscopy and 100 μm for brightfield microscopy.

At the next stage, in the animal group 4 the number of transplanted hMSCs was increased from $5x10^5$ hMSCs to $1x10^6$ hMSCs in 1 ml of PBS. The other injection parameters were left the same as for group 3. However, as can be seen from Figure 5, cerebral embolism was again observed in all rats in that group, suggesting that the usage of high doses of hMSCs in potentially dangerous even combined with slow infusion rate and maintenance of blood flow in ICA.

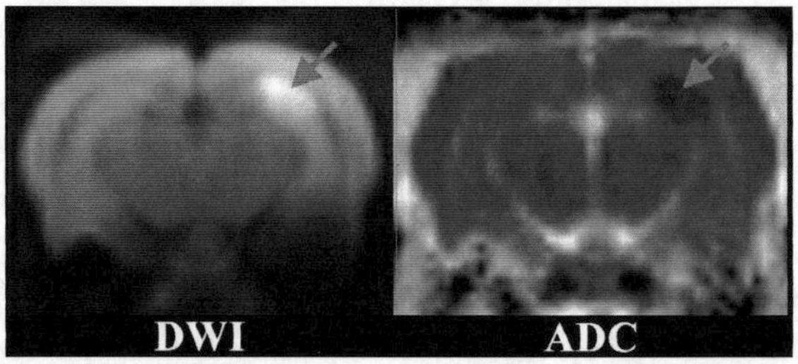

Figure 5. MRI of the brain of a rat from group 4 immediately after intra-arterial injection of $1x10^6$ SPIO-labeled hMSCs by syringe pump at the rate of 100 μ/min and maintenance of blood flow in the ICA. Red arrow points to an infarction zone formed due to cellular embolism.

3.2.2 Rats with MCAO. After determining optimal injection parameters for intact rats, we assessed their applicability in MCAO cerebral infarction model. In experimental group 5 24h after MCAO $1x10^5$ hMSCs in 1 ml of PBS were administrated by syringe pump with the speed of 100 μ/min and maintenance of blood flow in the ICA. The 24h time window appears to be the most reasonable according to the literature data and possible clinical application [4].

No cerebral embolic events were observed in the majority of rats (in 22 out of 27 rats). As seen in Figure 6, no increase of the infarction zone was detected. SPIO-labeled hMSCs in the right hemisphere were appropriately visualized with SWI as hypointense spots, which was confirmed by histology, as shown in Figure 4 (c).

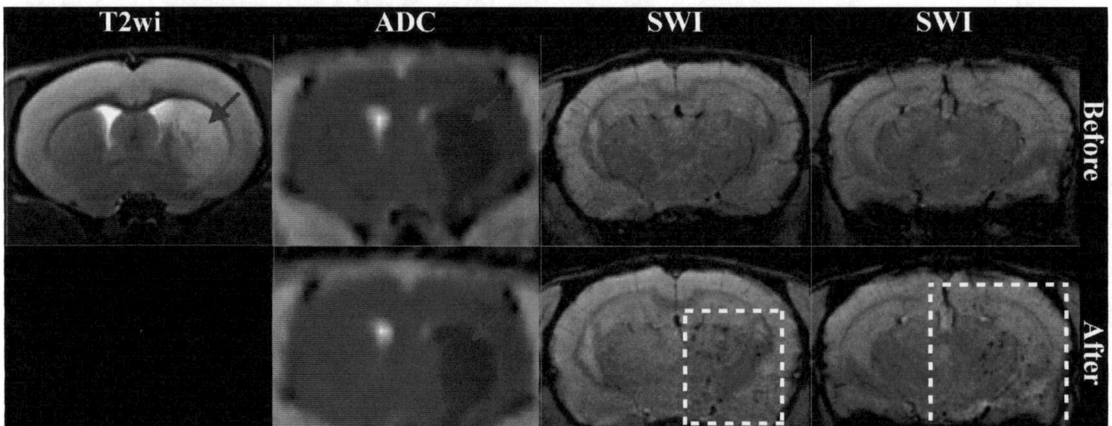

Figure 6. MRI of the brain of a rat from group 5 (transplantation of hMSCs 24h after MCAO, optimal parameters of cell administration) before and immediately after intra-arterial injection of SPIO-labeled hMSCs. Stroke zone remains unaltered after injection. Blue arrows indicate the infarction zone after MCAO. White dotted box indicates the area of hMSCs distribution in the right brain hemisphere. SPIO-labeled hMSCs are visualized as hypointense spots with SWI.

In 5 rats of this group the spread of the cerebral infarction zone after cells administration was revealed, suggesting cellular embolism (Figure 7). Embolic complications observed after the optimization of the parameters of intra-arterial transplantation can be the result of changes in intracerebral hemodynamics in the area around brain infarction or too deep insertion of the catheter inside ICA and lack of blood flow around it.

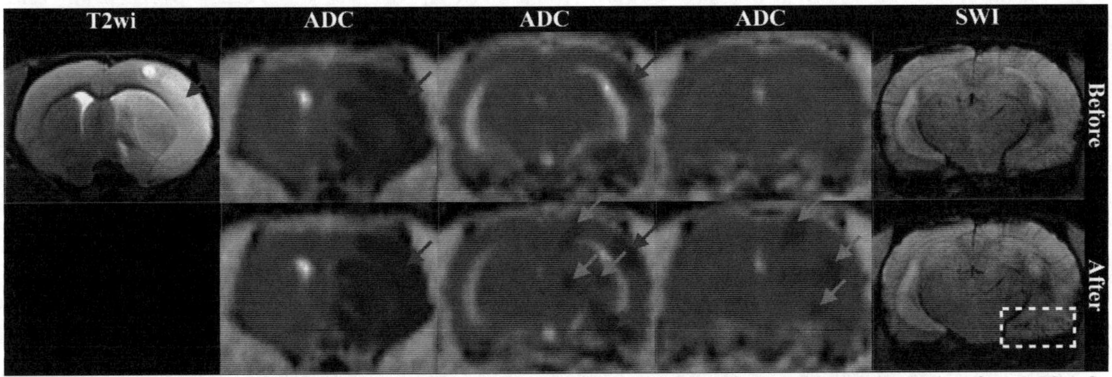

Figure 7. MRI of the brain of a rat from group 5 (transplantation of hMSCs 24h after MCAO, optimal parameters of cell administration) before and immediately after intra-arterial injection of SPIO-labeled hMSCs. New areas of ischemia due to cellular embolism can be observed. Blue arrows show the infarction zone after MCAO, red arrows indicate new infarction areas due to cellular embolism. White dotted box indicates the area of hMSCs distribution in the right brain hemisphere. SPIO-labeled hMSCs are visualized as hypointense spots with SWI.

Summarized data related to cerebral embolic complications after intra-arterial injection of SPIO-labeled hMSCs for all groups of rats are presented in Figure 8.

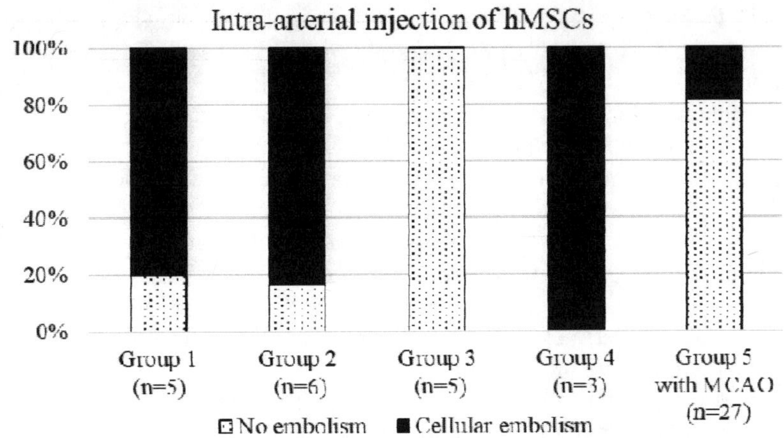

Figure 8. Percentage of rats with and without complications (cerebral embolism) after intra-arterial injection of SPIO-labeled hMSCs in different experimental groups. Detailed description of each group is given in the Materials and Methods section.

4. Conclusions

In summary, our results provide the optimal parameters for safety intra-arterial administration of hMSCs in intact rats and after MCAO. We have shown that the incidence of cerebral embolic events depends on such parameters as cell dose, infusion rate and maintenance of blood flow in ICA. Moreover, the correct ligation of intracerebral brunches of ICA can improve target cell delivery to the brain. Obtained data should be considered before planning experiments with intra-arterial administration of cells in laboratory animals and can help in further clinical trials with intra-arterial transplantations of stem cells in ischemic stroke patients.

Acknowledgements
The authors acknowledge the support by the Russian Foundation for Basic Research, grant No 16-29-07116 (cell culturing and labelling, animal surgery, MRI examination, histology and microscopy) and by the Russian Science Foundation, grant No 14-45-00040-П for the manuscript preparation and publication.

References
[1] Powers W, Derdeyn C, Biller J, Coffey C, Hoh B, Jauch E, Johnston K, Johnston S, Khalessi A, Kidwell C, Meschia J, Ovbiagele B and Yavagal D 2015 AHA/ASA Focused Update of the 2013 Guidelines for the Early Management of Patients With Acute Ischemic Stroke Regarding Endovascular Treatment *Stroke*
[2] Bang O, Kim E, Cha J and Moon G 2016 Adult Stem Cell Therapy for Stroke: Challenges and Progress. *J. stroke* **18** 256–66
[3] Kassis I, Petrou P, Vaknin-Dembinsky A and Karussis D 2011 *Clinical Investigation* vol **3** (Future Science)
[4] Toyoshima A, Yasuhara T, Kameda M, Morimoto J, Takeuchi H, Wang F, Sasaki T, Sasada S, Shinko A, Wakamori T, Okazaki M, Kondo A, Agari T, Borlongan C V and Date I 2015 Intra-Arterial Transplantation of Allogeneic Mesenchymal Stem Cells Mounts Neuroprotective Effects in a Transient Ischemic Stroke Model in Rats: Analyses of Therapeutic Time Window and Its Mechanisms. *PLoS One* **10** e0127302
[5] Misra V, Ritchie M, Stone L, Low W and Janardhan V 2012 Stem cell therapy in ischemic stroke: role of IV and intra-arterial therapy. *Neurology* **79** S207-12
[6] Boltze J, Arnold A, Walczak P, Jolkkonen J, Cui L and Wagner D-C 2015 The dark side of the force – constraints and complications of cell therapies for stroke *Front. Neurol.* **6** 155
[7] Cui L, Kerkelä E, Bakreen A, Nitzsche F, Andrzejewska A, Nowakowski A, Janowski M, Walczak P, Boltze J, Lukomska B and Jolkkonen J 2015 The cerebral embolism evoked by intra-arterial delivery of allogeneic bone marrow mesenchymal stem cells in rats is related to cell dose and infusion velocity. *Stem Cell Res. Ther.* **6** 11
[8] Fukuda Y, Horie N, Satoh K, Yamaguchi S, Morofuji Y, Hiu T, Izumo T, Hayashi K, Nishida N and Nagata I 2015 Intra-Arterial Transplantation of Low-Dose Stem Cells Provides Functional Recovery Without Adverse Effects After Stroke *Cell. Mol. Neurobiol.* **35** 399–406
[9] Longa E, Weinstein P, Carlson S and Cummins R 1989 Reversible middle cerebral artery occlusion without craniectomy in rats. *Stroke* **20** 84–91
[10] Koizumi J, Yoshida Y, Nakazawa T and Ooneda G 1986 Experimental studies of ischemic brain edema. I. A new experimental model of cerebral embolism in rats in which recirculation can be introduced in the ischemic area. *Jpn J Stroke* **8** 1–8
[11] Uluç K, Miranpuri A, Kujoth G, Aktüre E and Başkaya M 2011 Focal Cerebral Ischemia Model by Endovascular Suture Occlusion of the Middle Cerebral Artery in the Rat *J. Vis. Exp.* e1978–e1978

Oligodendrogenesis and neurogenesis in remyelination in the cuprizone model of multiple sclerosis: correlation with the degree of lesion

A Pishchelko[1], M Khodanovich[1], E Pan, V Glazacheva[1], A Akulov[1,2] and V Yarnykh[1,3]

[1] Research Institute of Biology and Biophysics, Tomsk State University, Tomsk, Russia
[2] Institute of Cytology and Genetics of the Russian Academy of Sciences, Novosibirsk, Russia
[3] Vascular Imaging Laboratory, Department of Radiology, University of Washington, Seattle, WA, USA

E-mail: bio_1979@mail.ru

Abstract. In this research, a cuprizone model of multiple sclerosis (MS) was used to study oligodendrogenesis and neurogenesis in remyelination. It has been shown that, with the administration of cuprizone, the amount of myelin in a number of structures of white and gray matter and the level of neurogenesis decrease, while the level of oligodendrogenesis increases. The withdrawal of cuprizone leads to the restoration of myelin content, the reduction of the excessive production of oligodendrocytes and to the restoration of the number of neurons to control values. The negative correlation between the number of oligodendrocyte precursors (OPCs) and the degree of demyelination of the corpus callosum indicates migration of OLG precursors from the subventricular zone (SVZ) to the structure during demyelination.

1. Introduction

One of the most significant demyelinating diseases, the development mechanisms of which still remain unclear, is multiple sclerosis (MS), which occurs mainly in young and middle-aged people and leads to a severe disability of patients. MS is characterized by a wave-like clinical course, during which exacerbations are replaced by remissions; therefore, the study of regeneration processes is extremely important for the disease prognosis, as well as for the development of new methods of the therapy.

In our study we resorted to a widely used cuprizone-induced model of MS, which allows studying the demyelination of brain nerve fibers that can be identified by both modern MRI methods and histology [1-3]. Administered orally for several weeks, the cuprizone toxin, being a copper chelator and an inhibitor of copper-dependent enzymes, leads to a selective death of oligodendrocytes (OLGs) and to the demyelination of the brain [4]. During several weeks following the withdrawal of cuprizone, an almost complete recovery of myelin is observed [5]. OPCs are the main population of cells involved in the recovery of myelin under various pathological conditions [6]. It is known that OPCs, at least in part, are formed in the SVZ, the so-called neurogenic niche, which is also one of the two known brain regions that generate neuronal precursors (NPCs) in adult age [7]. In a healthy brain, the neurons that form in the SVZ

IOP Conf. Series: Journal of Physics: Conf. Series **886** (2017) 012013 doi:10.1088/1742-6596/886/1/012013

migrate to the olfactory bulb, where they turn into adult interneurons [7,8]. OPCs formed in the SVZ normally migrate to the corpus callosum (CC) and cortex [6,8]. Neurogenesis also undergoes changes in the cuprizone model of demyelination [9]. We assumed that, during demyelination/remyelination, neurogenesis and oligodendrogenesis are inversely related, and the differentiation of stem cells shifts towards OPCs.

The aim of the work was to study the changes in neurogenesis and oligodendrogenesis during demyelination and remyelination, as well as studying their dependence on the extent of damage in the cuprizone model of MS.

2. Methods

Demyelination was induced in 4 male CD-1 mice by administering 0.5% cuprizone orally (mixed with standard food) for 10 weeks. To study remyelination, mice (N=4) were fed 0.5% cuprizone mixed with standard food for 5 weeks, after which they were returned to normal food. The control group of mice (N=4) received standard food. On the last day of the experiment, all the animals were subjected to an MRI imaging study, using a scanner for small laboratory animals Bruker "BioSpec 117/16USR" 11.7T under gas anesthesia with 1.5-2% isoflurane. The scanning was carried out in accordance with the protocol for obtaining T2-weighted images (100x100 μm resolution, 0.5 mm section thickness). Then the animals underwent a transcardial perfusion with 4% paraformaldehyde under a light ether anesthesia, and the brain was removed and frozen for a further immunohistochemical examination. Image processing was carried out in ImageJ software application. The T2-weighted images were used to assess the degree of lesion by checking the size of the corpus callosum, which was manually contoured in three consecutive sections (-0.82 ÷ -0.94 from bregma according to the atlas [10] (Figure 1). The sections were stained with the following markers: 1) neuro-glial antigen 2 (NG2) – a marker of young OLG precursors, 2) doublecortin (DCX) – a marker of immature neurons, and 3) a marker of the myelin basic protein (MBP). The mean intensity of MBP staining was measured in the following structures: corpus callosum, commissura anterior, capsula interna, thalami, caudoputamen, hippocampus, and cerebral cortex. Oligodendrogenesis was assessed in the same structures. Neurogenesis was assessed in typical neurogenic zones – lateral ventricles (SVZ) and dentate gyrus (DG) – by means of counting DCX-positive cells. The groups were compared using an independent Student test, and the correlation relationships were examined with the use of the Pearson correlation coefficient.

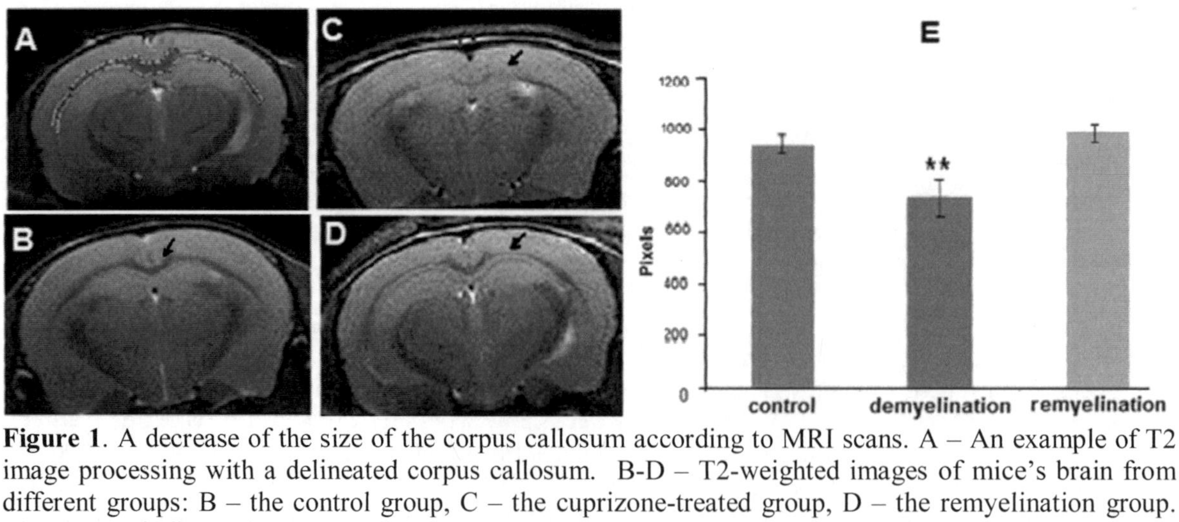

Figure 1. A decrease of the size of the corpus callosum according to MRI scans. A – An example of T2 image processing with a delineated corpus callosum. B-D – T2-weighted images of mice's brain from different groups: B – the control group, C – the cuprizone-treated group, D – the remyelination group. The arrows indicate the decreased corpus callosum in a mouse from the demyelination group and the normal size of the corpus callosum in mice from the control and remyelination groups. The significant decrease of the size of the corpus callosum in cuprizone-treated mice compared to the control group.** – $p<0.01$, T-test.

IOP Conf. Series: Journal of Physics: Conf. Series **886** (2017) 012013 doi:10.1088/1742-6596/886/1/012013

3. Results

The resulting T2 images clearly visualize differences between the groups in the degree of myelination in the brain structures of the mice. The changes in the corpus callosum are particularly noticeable: when cuprizone is introduced, the size of the corpus callosum becomes reduced by 19.5% as compared to the control group ($p<0.01$), while a recovery of the volume to the control level is observed at remyelination (Figure 1). The intensity of MBP staining in the demyelination group is lower than that in the control group in all brain structures ($p<0.05$) with the exception of the corpus callosum, front commissura and internal capsula (Figure 2).

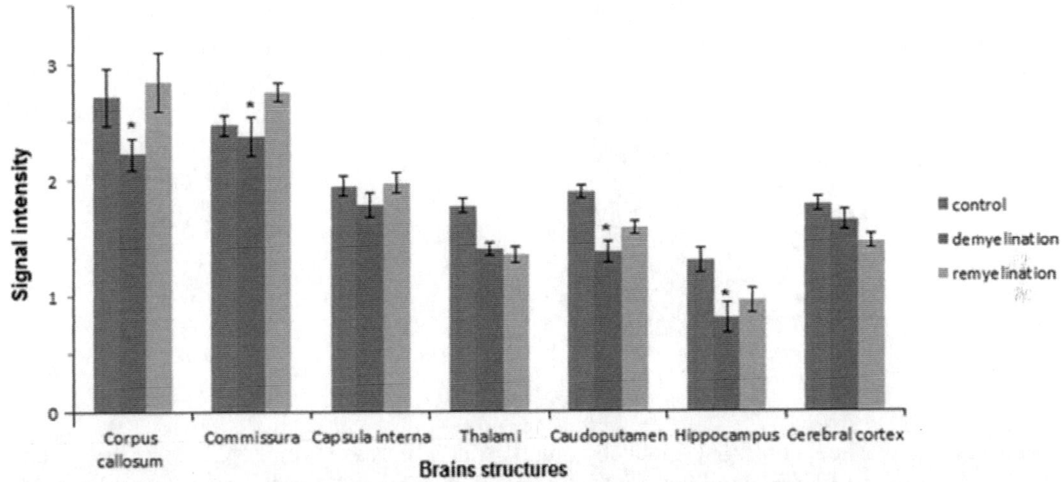

Figure 2. The significant decrease of myelin basic protein (MBP) in CPZ-treated mice compared to the control and remyelination groups. * – $p<0.05$, T-test.

At remyelination, the level of fluorescence signal on the stained MBP sections is restored to the control values. Thus, a noninvasive assessment of demyelination by T2-weighted images is confirmed by immunohistochemical staining data of the myelin basic protein. The DCX-staining revealed a significant reduction in neurogenesis in the demyelination group ($p<0.001$), whereas in the remyelination group the values did not differ from the control ones (Figure 3).

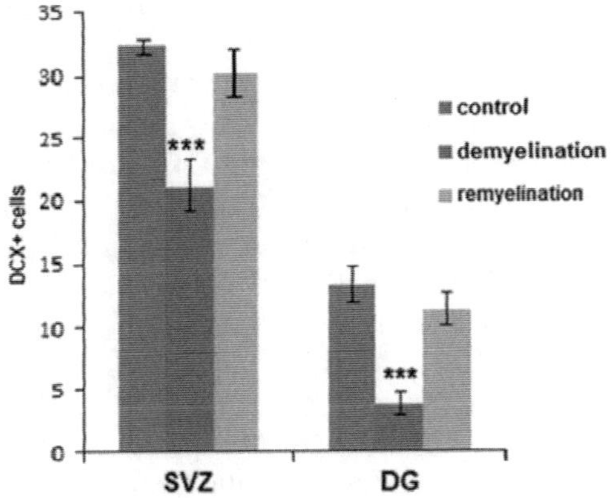

Figure 3. The cuprizone-induced significant decrease in DCX+ cells (newborn neurons) in the dentate gyrus (DG) and the subventricular zone (SVZ) was restored in the remyelination group. Significant differences compared with the control group: ** – $p<0.001$, T-test.

IOP Conf. Series: Journal of Physics: Conf. Series **886** (2017) 012013 doi:10.1088/1742-6596/886/1/012013

The NG2-staining, on the contrary, showed an increase in the level of oligodendrogenesis in the demyelination group for all investigated structures as compared to the control group ($p<0.001$), while there were no significant differences between the control group and the group of mice with remyelination (Figure 4). Moreover, a negative correlation was found between the number of NG2+ cells in the structure and the size of the corpus callosum in the resulting T2-weighted image ($r = -0.70$, $p<0.05$).

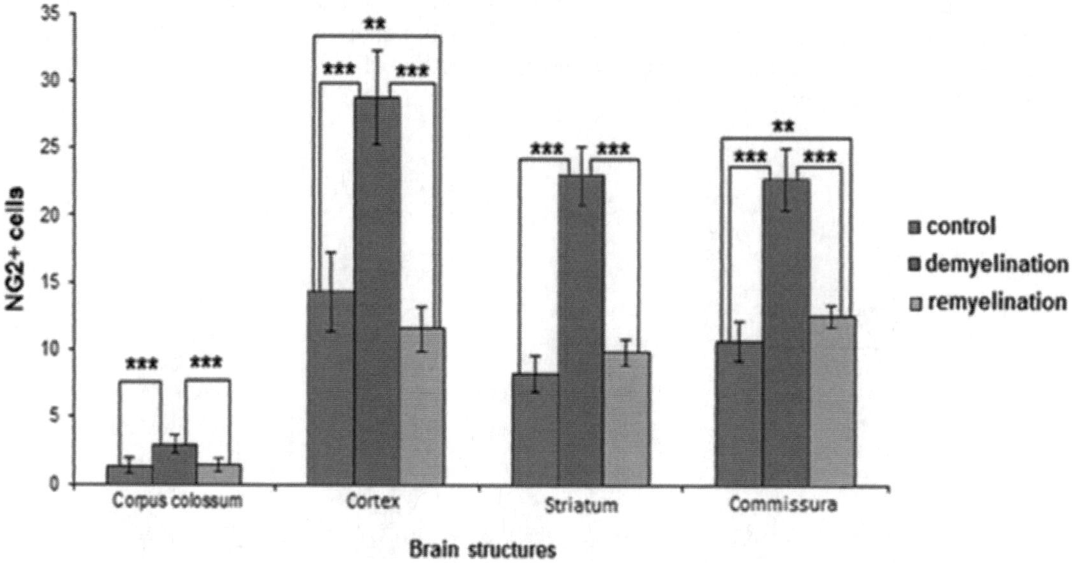

Figure 4. The cuprizone-induced significant increase in NG2+ cells (immature oligodendrocytes). The number of NG2+ cells is restored up to the control level in the remyelination group in the corpus callosum and striatum. The significant differences between groups: ** – $p<0.01$, * – $p<0.05$, T-test.

4. Discussion

It is known that cuprizone is a chelator of copper, and it leads to the inhibition of copper-dependent mitochondrial enzymes of cytochrome oxidase and monoamine oxidase. The administration of cuprizone causes the death of mature oligodendrocytes and demyelination, followed by remyelination upon the withdrawal of cuprizone [5,6]. The results of our experiments also indicate the loss of myelin during demyelination and its recovery during remyelination, which is shown by both MRI studies and immunohistochemical staining of the myelin basic protein.

As shown by the experimental data, oligodendrogenesis increases with demyelination and decreases with remyelination, which is consistent with the data of other researchers on the early OPC reaction after demyelination, where proliferation and colonization of the lesion area by OLG precursors was demonstrated [11-14]. This is due to the presence of OPCs, which are preserved in the process of cuprizone intoxication during demyelination since their metabolism is still not very active [15,16]. A part of OPCs migrates radially from the SVZ zone in the direction of the demyelinated corpus callosum. Next, OPCs generate OLGs. The newly formed OPCs and mature OLGs are much more exposed to the strong oxidative stress and the stress of the endoplasmic reticulum (ER), which leads to the death of OLGs [15-17]. The negative correlation between the degree of demyelination and the number of OPCs in the structure of the corpus callosum can indicate that the OPC cells migrating from the SVZ zone make up the majority of the OLG precursors at the time of demyelination since the migration path of these cells lies exactly in this direction. At the same time, the negative correlation between the number of OPC and the degree of lesion, as assessed by the size of the corpus callosum, indicates an intensification of the regeneration processes with a more pronounced intoxication with cuprizone. The destruction of the OLG perikaryon and of the myelin sheath causes apoptosis of OLGs after several days of the exposure to cuprizone, and on the fourth week there is a massive apoptosis of OLGs under the influence of the immune system. By the sixth week, a process of remyelination is observed, with OPCs actively differentiating into OLGs [18-24].

According to our results, during the process of demyelination, the level of neurogenesis in both neurogenic zones decreases. According to publications, neurodegeneration at cuprizone intoxication can occur as a result of the disruption of the metabolic connection between neurons and astrocytes [25]. At that, the destruction and extensive degeneration of the perikaryon of the neuron in the hippocampus area can reach up to 50%, but they can quickly recover afterwards, during the subsequent 5-week remyelination [26]. Neurodegeneration should lead to more intensive recovery processes, i.e. to the enhancement of neurogenesis in the proliferative zones; however, we observed a reverse picture in our experiment. Since the administration of cuprizone leads primarily to a selective death of OLGs, and a significant part of OPCs originates from stem cells in neurogenic niches, it is likely that the reduction of neurogenesis during demyelination and its recovery during remyelination is associated with a shift in the differentiation of stem cells towards OPCs.

5. Conclusion

This study has shown that, with the administration of cuprizone, the amount of myelin in a range of structures of white and gray matter and the level of neurogenesis decrease, while the level of oligodendrogenesis increases. The withdrawal of cuprizone leads to the restoration of myelin content, to the reduction of the excessive production of oligodendrocytes, and to the restoration of the number of neurons to control values. The negative correlation between the number of OPCs and the degree of demyelination of the corpus callosum indicates migration of OPCs from the subventricular zone (SVZ) to the demyelinating structures.

Acknowledgements

The study was supported by Russian Science Foundation, project No.14-45-00040-П.

References

[1] Wood T, Simmons C, Hurley S,. Vernon A, Torres J, Dell'Acqua F,. Williams S and Cash D. 2016 Whole-brain *ex-vivo* quantitative MRI of the cuprizone mouse. *PeerJ* **4** e2632.

[2] Khodanovich M, Sorokina I, Glazacheva V, Akulov A, Nemirovich-Danchenko N, Romashchenko A, Tolstikova T, Mustafina L and Yarnykh V 2017 Histological validation of fast macromolecular proton fraction mapping as a quantitative myelin imaging method in the cuprizone demyelination model. *Sci Rep.* **7** 46686

[3] Krutenkova E., Pan E., Khodanovich M. A 2015 Review of MRI Evaluation of Demyelination in Cuprizone Murine Model. *AIP Conference Proceedings* **1688** 030003

[4] Benetti F, Ventura M, Salmini B, Ceola S, Carbonera D, Mammi S, Zitolo A, D'Angelo P, Urso E, Maffia M, Salvato B, Spisni E 2010 Cuprizone neurotoxicity, copper deficiency and neurodegeneration *Neurotoxicology* **31** 509–517

[5] Zawadzka M, Rivers L E, Fancy S P, Zhao C,Tripathi R, Jamen F,Young K, Goncharevich A, Pohl H, Rizzi M, Rowitch D H, Kessaris N, Suter U, Richardson W D, Franklin R J 2010 CNS-Resident Glial Progenitor/Stem Cells Produce Schwann Cells as well as Oligodendrocytes during Repair of CNS Demyelination *Cell Stem Cell* **4** 578–590

[6] El Waly B, Macchi M, Cayre M, Durbec P 2014 Oligodendrogenesis in the normal and pathological central nervous system *Front Neurosci* **8** 145

[7] Lois C, Alvarez-Buylla A 1993 Proliferating subventricular zone cells in the adult mammalian forebrain can differentiate into neurons and glia *Proc Natl Acad Sci. U.S.A.* **90** 2074–2077

[8] Doetsch F, Caille I, Lim D A, Garcia-Verdugo J M, Alvarez-Buylla A 1999 Subventricular zone astrocytes are neural stem cells in the adult mammalian brain *Cell* **97** 703–716

[9] Khodanovich M, Glazacheva V, , Pan E, Akulov A, Krutenkova E, Trusov V and Yarnykh V. 2016 Polyprenols stimulate neurogenesis and reduce demyelination in the murine cuprizone model. *Annals of Neurology* **80(20)** S175

[10] Paxinos G, Frankin B J 2001 The mouse brain in stereotactic coordinates (San Diego, CA: Academic Press)

[11] Taylor L C, Puranam K, Gilmore W, Ting J P, Matsushima G K 2010b 17betaEstradiol protects male mice from cuprizone-induced demyelination and oligodendrocyte loss *Neurobiol Dis* **39** 127–137

[12] Venturini G 1973 Enzymic activities and sodium, potassium and copper concentrations in mouse brain and liver after cuprizone treatment in vivo *J Neurochem.* **21** 1147–1151

[13] Franklin R J, Gilson J M, Blakemore W F 1997 Local recruitment of remyelinating cells in the repair of demyelination in the central nervous system *J Neurosci* **50** 337–344

[14] Levine J M, Reynolds R 1999 Activation and proliferation of endogenous oligodendrocyte precursor cells during ethidium bromide-induced demyelination *Exp Neurol* **160** 333–347

[15] Blakemore W F, Smith P M, Franklin R J 2000 Remyelinating the demyelinated CNS *Novartis Found Symp* **231** 298–306

[16] Chari D M, Blakemore W F 2002 Efficient recolonisation of progenitor-depleted areas of the CNS by adult oligodendrocyte progenitor cells *Glia* **37** 307–313

[17] Bernardo A, Greco A, Levi G, Minghetti L 2003 Differential lipid peroxidation, *Mn* superoxide, and bcl-2 expression contribute to the maturation-dependent vulnerability of oligodendrocytes to oxidative stressJ. *Neuropathol Neurol* **62** 509–519

[18] Butts B D, Houde C, Mehmet H 2008 Maturation-dependent sensitivity of oligodendrocyte lineage cells to apoptosis: implications for normal development and disease *Cell Death Differ* **15** 1178–1186

[19] _Juurlink B H, Thorburne S K, Hertz L 1998 Peroxide-scavenging deficit underlies oligodendrocyte susceptibility to oxidative stress *Glia* **22** 371–378

[20] Dupree J L, Mason J L, Marcus J R, Stull M, Levinson R, Matsushima G K, Popko B 2004 Oligodendrocytes assist in the maintenance of sodium channel clusters independent of the myelin sheath *Neuron Glia Biol* **1** 179–192

[21] Goldberg J, Daniel M, van Heuvel Y, Victor M, Beyer C, Clarner T, Kipp M 2013 Short-term cuprizone feeding induces selective amino acid deprivation with concomitant activation of an integrated stress response in oligodendrocytes *Cell. Mol. Neurobiol* **33** 1087–1098

[22] Gudi V, Skuljec J, Yildiz O, Frichert K, Skripuletz T, Moharregh-Khiabani D, Voss E, Wissel K,Wolter S, Stangel M 2011 Spatial and temporal profiles of growth factor expression during CNS demyelination reveal the dynamics of repair priming *PLoS One* **6** e22623

[23] Hesse A, Wagner M, Held J, Bruck W, Salinas-Riester G, Hao Z, Waisman A, Kuhlmann T 2010 In toxic demyelination oligodendroglial cell death occurs early and is FAS independent *Neurobiol Dis* **37** 362–369

[24] Mason J L, Ye P, Suzuki K, D'Ercole A J, Matsushima G K 2000 Insulin-like growth factor-1 inhibits mature oligodendrocyte apoptosis during primary demyelination *J. Neurosci* **20** 5703–5708

[25] Morell P, Barrett C V, Mason J L, Toews A D, Hostettler J D, Knapp G W, Matsushima G K 1998 Gene expression in brain during cuprizone-induced demyelination and remyelination *Mol. Cell. Neurosci.* **12** 220–227

[26] Doan V, Kleindienst A M, McMahon E J, Long B R, Matsushima G K, Taylor L C 2013 Abbreviated exposure to cuprizone is sufficient to induce demyelination and oligodendrocyte loss *J Neurosci Res* **91** 363–373

Neuroimaging patterns of cerebral hyperperfusion

S Semenov[1], Yu Portnov[1], A Semenov[2], A Korotkevich[1] and A Kokov[1]

[1]Research Institute for Complex Issues of Cardiovascular Diseases, Kemerovo, Russia
[2]Praxis Wolfgang Theobald Facharzt für Radiologie, Saarlouis, Saarland, Deutschland

E-mail: dr_semenov_s@mail.rum

Abstract. Cerebral hyperperfusion syndrome (CHS) after revascularization is a rare phenomenon associated with post-ischemic (reactive) hyperemia and acute pathological hyperperfusion. First described on perfusion CT as a very often moderate CBF increase, MTT/TTP decrease within 30% like a temporary effect, according to a short-time deterioration of neurological symptoms (vestibular ataxia - 58%, vegetative dysfunction - 100%, asthenic syndrome - 100%) in early postoperative period in patients with cardiac ischemia who had undergone coronary artery bypass surgery. The acute pathological hyperperfusion carotid revascularization is a casuistic phenomenon with two- or three-fold CBV and MTT/TTP increase and high hemorrhage risk. Besides, we detected similar exchanges via perfusion CT called benign hyperemia, which marks extension of MTT/TTP and an increase of CBV from 27% to 48% (average 30%), but with normal CBF-parameters, indicating that venous stasis in acute venous ischemic stroke due cerebral venous sinus-trombosis (68%), only 6% in cardioembolic stroke and appears never in arterial stroke. Territorial coincidence registered for perifocal of necrosis zones of benign hyperemia and vasogenic edema accompanied on MRI (DWI, ADC). Secondary hemorrhagic transformation registered for primary non-hemorrhagic venous stroke in 27%, only in 9% for arterial stroke and in 60% for cardioembolic stroke. Probably, congestion is an increasingly predisposing factor secondary hemorrhaging than necrosis.

1. Introduction

Cerebral hyperperfusion syndrome (CHS) after cerebral revascularization is a well-described rare phenomenon after carotid endarterectomy, carotid artery stenting or after intravenous tissue plasminogen activation [1, 2]. There were described increased relative cerebral blood flow (CBF, ml/100g/min) and also relative cerebral blood volume (CBV, ml/100g); decreased mean transit time (MTT/TTP, s) like a temporary effect with recovery rates to normal and called the post-ischemic (reactive) hyperemia and acute pathological hyperperfusion - CHS [3, 4]. The most important pathophysiological factor in pathological hyperperfusion is apparently lowered cerebrovascular reserve as a result of prolonged and severe hypoperfusion [5]. There have been reports of cerebral perfusion studies with venous infarction in the experiments with ligation [6] and in the clinical practice as an example of a single event with dural sinus sagittalis superior thrombosis on the young men [7]. Acute venous stroke (VS) in contrast to the patterns with arterial acute ischemic stroke (AIS) is characterized by an increase in regional CBF and CBV when also the MTT/TTP is prolonged, indicating venous stasis [8]. A similar picture in the perfusion maps has areas of so-called benign hyperemia, which marks extension of MTT/TTP and an increase CBV, but normal CBF during perfusion tomographic studies [9].

2. Questions and objects of investigation

We did not meet the quantitative data on the deviation indicators CBF, CBV, MTT / TTP, which would characterize the boundary between the reactive hyperperfusion, CHS and benign hyperemia or contradictory opinions about the changes of MTT/TTP. The types of perfusion and diffusion of brain tissue disorders, were studied by us in patients with cerebral venous sinustrombosis (CVST) with the development of venous stroke without primary hemorrhage (VS, n=39), venous infarction due CVST with primary hemorrhage (n = 9), with atherothrombotic ischemic stroke (AIS, n = 56) and cardioembolic stroke (CS, n = 46), as well as in patients with ischemic heart disease with arterial hypertension, underwent coronary artery bypass grafting (n=47).

3. Benign hyperemia

Areas of hypoperfusion (penumbra) and aperfusion (necrosis/infarction) in patients with primary non-haemorrhagic stroke of AIS and CS on CBV maps with CTP, done within the therapeutic window is almost always accurately territorially correlated (r = 0.97) with signs of cytotoxic edema on diffusion-weighted images (DWI) and maps of the apparent diffusion coefficient (ADC) with magnetic resonance imaging (MRI). Territorial coincidence is registered for zones benign hyperemia and edema according to DWI, ADC at VS (Figure 1). Signs of benign hyperemia we met more often at the VS, but also, though rarely with CS. In perifocal zone of infarction we observed as a congestion on the map at CBV of CTP and images non-contrast MR perfusion (ASL). A distinctive feature of venous stroke from arterial lesions is the less frequent development of necrosis (55% vs. 79%), and also not oligemia in the perifocal infarction zone, but benign hyperemia with exceeding perfusion parameters compared with a healthy tissue on the opposite side from 27% to 48% (An average of 30%).

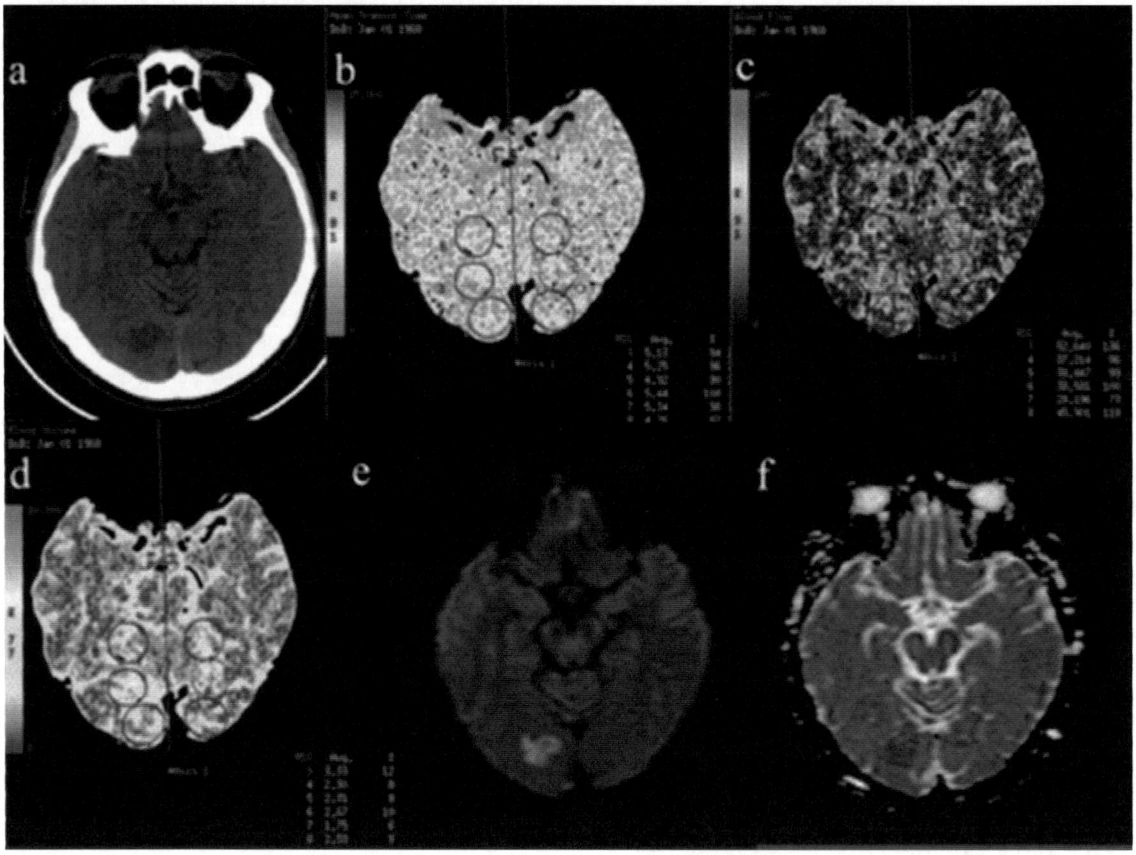

Figure 1. Hypodense right occipital zone on the initial CT in acute right transversal dural sinus thrombosis case (a). Congestive hyperperfusion on computed tomography perfusion: increased up to 25-30% mean transit time (b); (c) cerebral blood flow (c); cerebral blood volume (d). Diffusion-weighted magnetic resonance imaging (e) and apparent diffusion coefficient (f) on MRI confirms focal acute ischemia territory of right occipital zone.

IOP Conf. Series: Journal of Physics: Conf. Series **886** (2017) 012014 doi:10.1088/1742-6596/886/1/012014

In these areas, we noted an inversion of MR signal on DWI and ADC-mapping, indicating the early development of vasogenic edema in the background of cytotoxic edema in the hyperperfusion areas [10].

We [11] agree with opinion [12], that the phenomenon of hyperperfusion is associated with the development of vasogenic edema, which by venous stroke often comes within 24 hours, while by arterial stroke associated with the subacute period (from 3 to 7 days). "Crosshair" curves signal DWI and ADC seems to AIS usually in 5-7 days period, when there is a vasogenic edema. The signal on DWI begins to decline, while ADC-mapping, on the contrary - to rise. For venous ischemia zone crosshairs DWI and ADC curves shifted closer to the clinic manifestation (for 1 day). If the projection VS earlier inversion of foci is observed in 68% of cases, a similar inversion in arterial stroke is marked only in 6% at CS.

4. Acute cerebral hyperperfusion syndrome

Model CHS may be hemorrhagic venous infarction resulting CVST. When CTP notes almost two- or even threefold increased CBV in the affected area with a slight increase of CBF due to two- or threefold prolonged MTT/TTP (Figure 2).

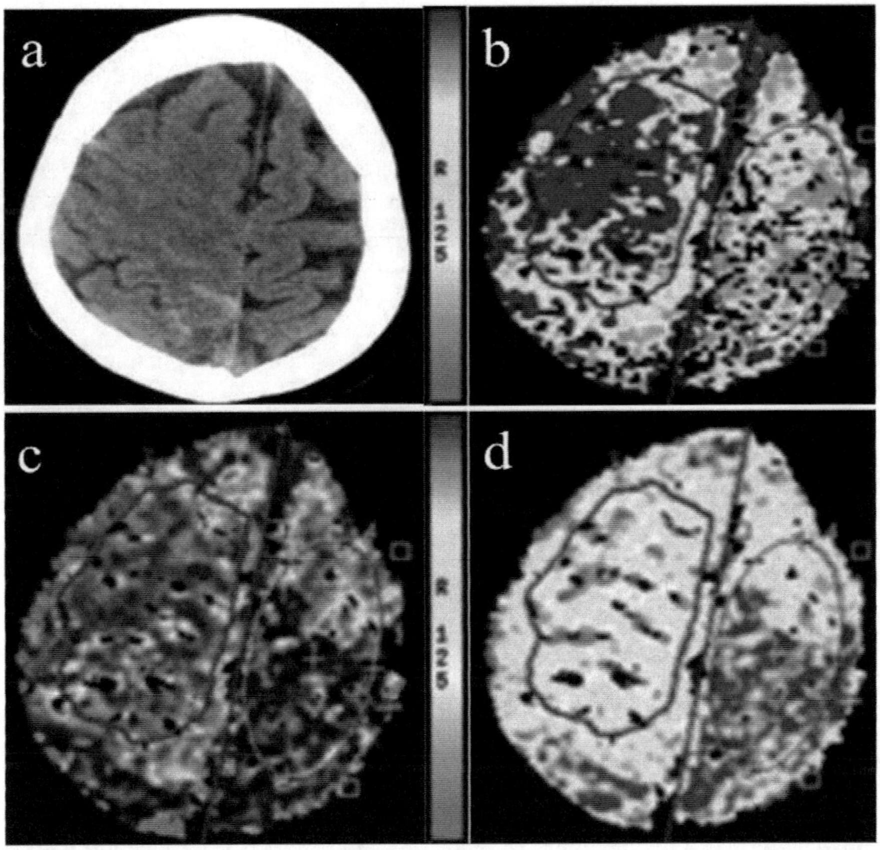

Figure 2. Edema of the superior parietal lobule and the SAH on the rightside on routine CT (a) in thrombosis superior sagittal sinus case. CTP marked almost three times larger of blood volume in the affected area (d) in an equally pronounced elongation contrast transit time (b), but a slight increase in blood flow (c).

Secondary hemorrhagic transformation registered on a control routine CT during the primary non-hemorrhagic VS 27%, only in 9% for AIS and in 60% for CS. A distinctive feature of secondary hemorrhage when VS is the development of congestion in the area, combined with the difficulty of outflow at vasogenic edema area. Hemorrhagic transformation at AIS develops in the area of necrosis and cytotoxic edema and it happens rarely. Probably, congestion is increasingly predisposing factor secondary hemorrhage than a necrosis. When CS combination of necrosis and hyperemia with elements of the early development of vasogenic edema in perifocal zones leads to secondary hemorrhage more frequently than in VS and AIS. Why is this happening? Since the constrictor response of the arteries is a compensatory protective nature, in its absence or reduction of the capacity for such a reaction in violation of cerebral haemodynamic autoregulation, the blood continues to flow in the cerebral vessels under ordinary or elevated pressure and hyperemia zone forms background of venous congestion. This results in an overflow of intracranial microcirculation, rise in intracranial pressure and brain-threatening consequences.

5. Reactive (post-ischemic) hyperperfusion

The phenomenon of reactive (post-ischemic) hyperperfusion observed by us in the early postoperative period in patients with cardiac ischemia and arterial hypertension with the duration more than 10 years who had undergone coronary artery bypass surgery (n = 47). There has been noted a moderate (up to 22-25%), same as significant change of cerebral perfusion (decrease of TTP, CBF and an increase of CBV) in areas perfused MCA on both sides. These changes were accompanied by a temporary deterioration of neurological symptoms: vestibular ataxia syndromes (58%), vegetative dysfunction syndrome (100%), asthenic syndrome (100%). Clinical manifestations of reperfusion syndrome in 8 days after the operation is not detected, whereas signs of cerebral hyperperfusion still persisted. Patients with an anamnesis of arterial hypertension (duration less than 10 years) or without hypertension significant changes in brain perfusion in the postoperative period were not found, which indicates the safety of cerebral haemodynamic autoregulation mechanisms.

6. Conclusion

In our opinion, changes TTP/MTT is not less informative data than the CBF and CBV (Table 1).

Table 1. Table 1. Type of focal hyperperfusion disorders of brain tissue.

Type of disorders of brain tissue perfusion	rMTT/TTP	rCBF	rCBV
Cerebral hyperperfusion syndrome	++	±	++
Benign hyperemia	±	±	+
Reactive postischemic hyperperfusion	-	+	±

Extension of time with a decrease in blood flow - a sign of hypoperfusion. A moderate increase in the time at moderately elevated blood volume - a sign of benign hyperemia. The shortening of the time with a moderate increase in blood volume - a sign of post-ischemic reactive hyperperfusion. Significant prolongation of transit time combined with a significant increase in blood volume are signs of cerebral hyperperfusion syndrome according with a high risk of hemorrhage. Close to each other benign hyperemia and reactive post-ischemic hyperperfusion have different mean transit time or time to peak (a small increase or a slight shortening, respectively).

Acknowledgmets
The authors acknowledge Russian Science Foundation, project No.14-45-00040-П, for manuscript preparation and publication.

References

[1] Seners P, Turc G, Tisserand M, Legrand L, Labeyrie M, Calvet D and Baron J 2014 Unex plained early neurological deterioration after intravenous thrombolysis incidence, predictors, and associated factors *J. Stroke* **45(7)** 2004-9

[2] Zhang Y, Kumar A, Tezel J B and Zhou Y 2016 Imaging evidence for cerebral hyperperfusion-syndrome after intravenous tissue plasminogen activator for acute ischemic stroke *J. Case Rep Neurol Med* **Epub 2016** 5 http://dx.doi.org/10.1155/2016/8725494

[3] Bazilevich S N, Odinak M M, Dyskin D E, Krasakov I V, Fokin V A, P'ianov I V, Dekan V S, Okol'zin A V and Stanzhevskiĭ A A 2008 The structural and functional neurovisualization in patients with epileptic seizures in cerebrovascular diseases *Zh Nevrol psikhiatr im S S Korsakova* **2** 33-9

[4] Nguyen T B, Lum C, Eastwood J D, Stys P K, Hogan M and Goyal M 2005 Hyperperfusion on perfusion computed tomography following revascularization for acute stroke *Acta Radiol* **46(6)** 610-5

[5] Hines G L, Oleske A, Feuerman M 2011 Post-carotid endarterectomy hyperperfusion syndrome. Is it predictable by lack of cerebral reserve? *Ann Vasc Surg* **25(4)** 502-7

[6] Schaller B, Graf R, Sanada Y, Tolnay M, Rosner G, Wienhard K and Heiss W 2003 Hemodynamic changes after occlusion of the posterior superior sagittal sinus: an experimental PET study in cats *AJNR* **24(9)** 1876-80

[7] Kawai N, Shindou A, Masada T, Tamiya T and Nagao S 2005 Hemodynamic and metabolic changes in a patient with cerebral venous sinus thrombosis: evaluation using O-15 positron emission tomography *Clin Nucl Med* **30(6)** 391-4

[8] Mullins M E, Grant P E, Wang B, Gonzalez R G, Schaefer P W 2004 Parenchymal abnormalities associated with cerebral venous sinus thrombosis: assessment with diffusion-weighted MR imaging *AJNR* **25(10)** 1666-75

[9] Adamczyk P, Liebeskind D S Topic review MRI perfusion imaging in acute ischemic stroke 2016 *University of California, Los Angeles* IndaBook.org © http://www.indabook.org

[10] Semenov S, Moldavskaya I, Shatokhina M, Semenov A, Barbarash L 2011 How to distinguish between venous and arterial strokes and why? *J. Neuroradiol* **24(2)** 289-99

[11] Semenov S, Moldavskaya I, Shatokhina M, Nekhorosheva A, Khromov A, Zhuchkova E et al 2012 CT and MRI patterns of focal hyperemia in venous insult *J. Neuroradiol* **54(1)** 176

[12] Gonzalez R, Hirsch J, Koroshetz W, Lev M, Schaefer P 2007 Acute ischemic stroke: imaging and intervention *AJNR* **28(8)** 1622

Novel targets for positron emission tomography (PET) radiopharmaceutical tracers for visualization of neuroinflammation

I Shchepetkin[1,2], M Shvedova[2,3], Y Anfinogenova[2,4], M Litvak[2,4,5] and D Atochin[2,5,6]

[1]Department of Microbiology and Immunology, Montana State University, Bozeman, Montana, USA
[2]RASA Center in Tomsk, Tomsk Polytechnic University, Tomsk, Russia
[3]Siberian State Medical University, Tomsk, Russia
[4]Research Institute for Cardiology, Tomsk NRMC, Tomsk, Russia
[5]Cardiovascular Research Center and Cardiology Division, Massachusetts General Hospital, Harvard Medical School, Charlestown, Massachusetts, USA
[6] Research Institute of Biology and Biophysics, Tomsk State University, Tomsk, Russia

E-mail: atochin@yahoo.com

Abstract. Non-invasive molecular imaging techniques can enhance diagnosis of neurological diseases to achieve their successful treatment. Positron emission tomography (PET) imaging can identify activated microglia and provide detailed functional information based on molecular biology. This imaging modality is based on detection of isotope labeled tracers, which emit positrons. The review summarizes the developments of various radiolabeled ligands for PET imaging of neuroinflammation.

Inflammation is involved in a variety of neurological diseases including stroke, Alzheimer disease, multiple sclerosis, Parkinson's disease, neurodegenerative dementias, epilepsy, psychiatric disorders such as schizophrenia, and oncologic diseases [1-6]. The inflammatory during most chronic neurodegenerative disease is dominated by the microglia [7]. Microglia are brain macrophages that emerge from early erythro-myeloid precursors and migrate to the brain mesenchyme before the blood brain barrier is formed [8]. Microglia constitute up to 10% of the total cell population of the brain. As resident macrophages of the central nervous system (CNS), the cells of microglia phagocytose cellular debris, present foreign antigens, and are the sensors of pathological events including inflammation [9]. Microglial cells can upregulate synthesis and release of various mediators, including translocator protein (TSPO), chemokines (e.g. CCL2), cyclooxygenase 1 (COX1), and cannabinoid receptor 2 (CB2) in the presence of inflammation. Blood-borne leukocytes including monocytes/macrophages, neutrophils, T-lymphocytes and B-lymphocytes extravasate into the brain through the interaction of cell surface integrins with specific endothelial adhesion molecules (e.g., ICAM-1, VCAM-1, P-/E-selectins). Stimulated cells then secrete effector molecules (e.g., matrix metalloproteinases [MMPs] and myeloperoxidase [MPO]), which trigger axonal damage and/or demyelination. Cell-to-cell interaction between the antigen-presenting cells (APCs, e.g., B-lymphocytes, microglia, dendritic cells) is mediated via CD40 among other molecules [10].

To date, the most promising imaging techniques to visualize neuroinflammation include the methods of positron emission tomography (PET) and single photon emission computed tomography (SPECT). Identification of neuroinflammation markers using PET is a promising area of research aimed to improve the accuracy of diagnosis of these diseases, specify the prognosis and assess the effects of the therapeutic interventions [4]. PET is an imaging technology that assays the distribution of ligands labelled with positron emitters (e.g., 18F, 11C, 15O) in vivo, by measuring the emitting annihilation photons with a ring of detectors. It is a functional imaging technology that best suits the three principles of tracer measurements because PET instrumentation (1) allows absolute quantification, (2) PET labels do not modify tracer properties, and (3) PET ligands are administered at trace concentrations. These properties stem from the unique ability of the technology to utilize high energy radiation emitted from the nucleus as opposed to lower energy modalities, such as computed tomography, where X-rays are emitted by the orbiting electrons. PET labels radiate positrons, the anti-matter counterpart of the electron that annihilate with slowly moving electrons emitting two γ-rays at an 180o angle that are detected by the ring of detectors. Once a study is completed, the coincident detections of the γ-rays are reconstructed into a set of time frames that measure the distribution of the radioligand during the acquisition [11].

Molecular imaging techniques that non-invasively visualize specific targets of the inflammation cascade using sensitive and specific probes may be powerful tools to evaluate neuroinflammation in the pre-clinical and clinical settings [12]. To date, a range of novel radioligand tracers for microglia inflammatory activation is under development. Imaging with 18F-fluorodesoxyglucose (FDG) PET is used to determine sites of abnormal glucose metabolism and can be used, for example, to characterize and to localize many types of tumors [13]. However, lack of specificity of 18F-FDG PET limits its application for visualization of sites of infection and inflammation.

Several radioligands have been developed to image the activation of microglia in experimental models and in various CNS diseases using PET techniques [14]. For example, translocator protein (TSPO) is consistently elevated in activated microglia of the CNS in response to a variety of adverse factors as well as in the presence of neurodegenerative and psychiatric conditions. 11C-PK11195 PET, aimed at imaging the expression of the TSPO on activated microglia in the brain, has been used in preclinical and clinical research to investigate neuroinflammation in vivo in patients with brain diseases. Increased 11C-PK11195 uptake has been observed around the ischemic lesions for several days after the onset, but also in the regions distant from the lesion [14]. However, 11C-PK11195 suffers from two major limitations: low brain permeability and high nonspecific signal due to plasma binding resulting in a low signal-to-noise ratio [4].

Significance of brain immunity as a primary or comorbid factor of illness has sparked great interest in the TSPO as a biomarker. In recent years, novel tracers such as [11C]-DAA1106, [11C]-vinpocetine, [11C]-DPA-713, [11C]-PBR28, [18F]-FEDAA1106, [18F]-PBR06, [18F]-PBR111, [18F]-DPA-714, and [18F]-FEPPA have been developed as the second-generation TSPO radioligands aimed at improving the quality of TSPO imaging through the higher affinity of novel radioligands [2, 15]. However, such major developments have not yet resulted in the expected improvement of image quality [5]. There are some challenges: for example, the second-generation TSPO ligands show high level of vascular binding, which makes it difficult to separate the normal gray matter from other tissues [16].

Apart from TSPO, other targets, involved in the neuroinflammatory process in neurodegenerative disease, have been identified and currently are evaluated using PET ligands. One promising target is the cannabinoid receptor type 2 (CB2R), which is involved in peripheral immune system function, and is also upregulated in CNS disorders showing microglial activation [17]. This has led to the development of specific ligands, such as [11C]-NE40, which show binding to the human CB2R overexpressed in rodent striatum in preclinical PET studies [18]. First-in-man study of [11C]-NE40 showed a swift brain uptake [19], however, at this time, no clinical data on CB2R PET studies in neuronal disease have been published yet [6].

PET tracers like [64Cu]DOTA-etanercept and [64Cu]pegylated dimeric c(RGDyK) have been developed to target tumor necrosis factor (TNF) in both acute and chronic inflammation in mice. TNF may be a target for multiple sclerosis imaging in the future [10]. Another marker participating in the neuronal damage is inducible nitric oxide (NO) synthase (iNOS). The number of tracers for iNOS is minimal and the current [18F]NOS (6-(1/2)(2-[18F]fluoropropyl)-4-methylpyridin-2-amine) requires further modification and improvement [10]. The expression of another proinflammatory cytokine mediator, cyclooxygenase-2(COX-2), is extensively increased in multiple sclerosis lesions and it has been tightly linked to increase in iNOS expression [20]. PET tracers for COX-2 have been developed, but the in vivo imaging properties have not been very effective [21]. The most promising COX-2 tracer so far is [11C]Rofecoxib (4-(4-methylsulfonylphenyl)-3-phenyl-5H-furan-2-one) demonstrating in vitro usability, but lacking necessary affinity for in vivo studies [22].

Formyl peptide receptors (FPRs) are the G protein-coupled receptors (GPCR) that play an important role in leukocyte activation and chemotaxis [23]. These receptors were originally identified for their ability to bind and be stimulated by N-formyl peptides, which are produced by bacteria but can also be released from damaged mitochondria during tissue injury. The FPR are abundantly expressed on the surface of macrophages and neutrophils [23]. Moreover, FPR2 could represent an interesting target for monitoring in vivo the onset and progression of neuroinflammation in Alzheimer disease [24]. In recent years, 64Cu isotopes have been linked to FPR ligands for preclinical and clinical imaging of inflammation sites by PET. The FPR-specific peptide, cinnamoyl-F-(D)L-F-(D)L-FK (cFLFLF), was sequentially conjugated with a bifunctional polyethylene glycol moiety (PEG) and a 2,2',2'',2'''-(1,4,7,10-tetraazacyclododecane-1,4,7,10-tetrayl)tetraacetic acid (DOTA) through a lysine (K) spacer and finally labeled with 64Cu2+ to form cFLFLFK-PEG-64Cu [25]. In vitro assays for binding of the synthesized ligand to the neutrophil FPR yielded a dissociation constant of 17.7 nM. PET imaging with cFLFLFK-PEG-64Cu revealed that lung standardized uptake values correlated with lung neutrophil activity in Klebsiella-infected mice [25]. This construct represents more effective PET agent for macrophage detection compared with 18F-FDG for detection of macrophages in the pancreatic islets and aorta [26]. However, the in vivo use of peptides for neurovisualization is hampered by their low ability to cross blood–brain barrier (BBB) unless they interact with specific transport systems. Thus, non-peptidic small molecules may have some advantages because they can be suitably designed to modulate properties, such as potency, selectivity, lipophilicity, and cell permeability, which are pivotal for a potential radiotracer [27]. Recently, a series of non-peptidic potent agonists for FPR1/2 with 3-(1H-indol-3-yl)-2-[3-(4-nitrophenyl)ureido]propanamide structure was reported [28]. On the basis this scaffold compound (S)-1 was selected as a candidate for visualization of FPRs in the activated microglial cells in vivo [27]. However, (S)-[11C]-1 showed very poor BBB penetration and, thus, was unable to accumulate in the brain [27].

Therefore, at the present time, searching for new neuroinflammation markers for PET detection remains a relevant problem and requires further research efforts.

Acknowledgments

The authors acknowledge the support from the Russian Science Foundation grant No17-15-01111 for manuscript preparation and publication.

References

[1] Heiss W-D 2014 PET imaging in ischemic cerebrovascular disease: current status and future directions *Neurosci Bull* **30** 713–32
[2] Zhang J 2015 Mapping neuroinflammation in frontotemporal dementia with molecular PET imaging *J Neuroinflammation* **12** 108
[3] Galovic M and Koepp M 2016 Advances of molecular imaging in epilepsy *Curr Neurol Neurosci* Rep. **16** 58

[4] Vivash L and O'Brien T 2016 Imaging Microglial Activation with TSPO PET: Lighting Up Neurologic Diseases. *J Nuc Med* **57** 165–8

[5] Turkheimer F E, Rizzo G, Bloomfield P S, Howes O, Zanotti-Fregonara P, Bertoldo A andVeronese M 2015 The methodology of TSPO imaging with positron emission tomography *Biochem Soc Trans* **43** 586–92

[6] Gerhard A 2016 TSPO imaging in parkinsonian disorders Clin. Transl. Imaging **4** 183–90

[7] Cunningham C 2013 Microglia and neurodegeneration: the role of systemic inflammation. *Glia* **61** 71-90

[8] Waisman A, Ginhoux F, Greter M and Bruttger J 2015 Homeostasis of Microglia in the Adult Brain: Review of Novel Microglia Depletion Systems. Trends Immunol. **36** 625-36

[9] Hanisch U K and Kettenmann H 2007 Microglia: active sensor and versatile effector cells in the normal and pathologic brain *Nat Neurosci* **10** 1387–94

[10] Poutiainen P, Jaronen M, Quintana F J and Brownell A-L 2016 Precision Medicine in Multiple Sclerosis: Future of PET Imaging of Inflammation and Reactive Astrocytes Front. *Mol Neurosci* **9** 85

[11] Beadsmoore C, Newman D, MacIver D and Pawaroo D 2015 Positron Emission Tomography Computed Tomography: A Guide for the General Radiologist Can. Assoc. *Radiol J* **66** 332–347

[12] Pulli B and Chen J W 2014 Imaging Neuroinflammation - from Bench to Bedside *J Clin Cell Immunol* **5** 226

[13] Wu C, Li F, Niu G and Chen X 2013 PET imaging of inflammation biomarkers *Theranostics* **3** 448-66.

[14] Thiel A and Heiss W D 2011 Imaging of microglia activation in stroke Stroke 42 507–12

[15] Buck A and Decristoforo C 2016 Highlights lecture EANM medicine's superheroes *Eur J Nucl Med Mol Imaging* **43** 1910–27

[16] Rizzo G, Veronese M, Tonietto M, Zanotti-Fregonara P, Turkheimer F and Bertoldo A 2014 Kinetic modeling without accounting for the vascular component impairs the quantification of [(11)C]PBR28 brain PET data *J Cereb Blood Flow Metab* **34** 1060–69

[17] Cabral G , Raborn E , Griffin L, Dennis J and Marciano-Cabral F 2008 CB2 receptors in the brain: role in immune function *Br J Pharmacol* **153** 240–51

[18] Evens N, Vandeputte C, Coolen C, Janssen P, Sciot R, Baekelandt V, Verbruggen A M, Debyser Z, Van Laere K and Bormans G M 2012 Preclinical evaluation of [11C]NE40, 2 cannabinoid receptor PET tracer *Nucl Med Biol* **39** 389–99

[19] Ahmad R, Koole M, Evens N, Serdons K, Verbruggen A, Bormans G and Van Laere K 2013 Whole-body biodistribution and radiation dosimetry of the cannabinoid type 2 receptor ligand NE40 in healthy subjects *Mol Imaging Biol* **15** 384–90

[20] Rose J W, Hill K E, Watt H E and Carlson N G 2004 Inflammatory cell expression of cyclooxygenase-2 in the multiple sclerosis lesion *J Neuroimmunol* **149** 40–9

[21] de Vries E F J , Doorduin, J, Dierckx R A and van Waarde A 2008 Evaluation of [(11)C]rofecoxib as PET tracer for cyclooxygenase 2 overexpression in rat models of inflammation *Nucl Med Biol* **35** 35–42

[22] Ji B, Kumata K , Onoe H , Kaneko H , Zhang M-R, Seki C, Ono M, Shukuri M, Tokunaga M, Minamihisamatsu T, Suhara T and Higuchi M 2013 Assessment of radioligands for PET imaging of cyclooxygenase-2 in an ischemic neuronal injury model *Brain Res* **1533** 152–62

[23] Ye RD, Boulay F, Wang JM, Dahlgren C, Gerard C, Parmentier M, Serhan CN and Murphy PM 2009 International Union of Basic and Clinical Pharmacology. LXXIII. Nomenclature for the formyl peptide receptor (FPR) family. *Pharmacol Rev* **61** 119-61

[24] Slowik A, Merres J, Elfgen A, Jansen S, Mohr F, Wruck C, Pufe T and Brandenburg LO 2012 Involvement of formyl peptide receptors in receptor for advanced glycation end products (RAGE) and amyloid beta 1-42-induced signal transduction in glial cells. *Mol Neurodegener* **7** 55

IOP Conf. Series: Journal of Physics: Conf. Series **886** (2017) 012015 doi:10.1088/1742-6596/886/1/012015

[25] Locke L W, Chordia M D, Zhang Y, Kundu B, Kennedy D, Landseadel J, Xiao L, Fairchild KD, Berr SS, Linden J and Pan D 2009 A novel neutrophil-specific PET imaging agent: cFLFLFK-PEG-64Cu 2009 J. Nucl. Med. **50** 790–97

[26] Zhang Y, Kundu B, Zhong M, Huang T, Li J, Chordia M, Chen M, Pan D, He J and Shi W 2015 PET imaging detection of macrophages with a formyl peptide receptor antagonist. Nucl Med Biol. **42** 381-6

[27] Lacivita E, Stama M, Maeda J, Fujinaga M, Hatori A, Zhang MR, Colabufo N, Perrone R, Higuchi M, Suhara T and Leopoldo M 2016 Radiosynthesis and in vivo Evaluation of Carbon-11 (2S)-3-(1H-Indol-3-yl)-2-{[(4-methoxyphenyl)carbamoyl]amino}-N-{[1-(5-methoxypyridin-2-yl)cyclohexyl]methyl}propanamide: An Attempt to Visualize Brain Formyl Peptide Receptors in Mouse Models of Neuroinflammation. Chem Biodivers. **13** 875-83

[28] Schepetkin I A, Kirpotina L N, Khlebnikov A I, Leopoldo M, Lucente E, Lacivita E, De Giorgio P and Quinn MT 2013 3-(1H-indol-3-yl)-2-[3-(4-nitrophenyl)ureido]propanamide enantiomers with human formyl-peptide receptor agonist activity: molecular modeling of chiral recognition by FPR2. Biochem. Pharmacol. **85** 404-16

The cerebral network's reconstruction by MRI methods and the hemodynamics study of small laboratory animal in type 1 diabetes

A Akulov[1,2] , A Cherevko[3,4], D Parshin[3], D Tur[1,4] and G Yankova[3,4]

[1]Institute of Cytology and Genetics of the Russian Academy of Sciences, Novosibirsk, Russia
[2]Research Institute of Biology and Biophysics, Tomsk State University, Tomsk, Russia
[3]Lavrentyev Institute of Hydrodynamics of the Russian Academy of Sciences, Novosibirsk, Russia
[4]Novosibirsk State University, Novosibirsk, Russia

E-mail: cherevko@inbox.ru

Abstract. The blood realizes the transport of substances, which are necessary for livelihoods, throughout the body. The assumption about the relationship some disease and structure of vasculature (in particular of brain) is natural. In the paper we consider models of Willis' circle for two groups of laboratory mice – one control group and another with diabetes. Vascular net obtained as a result of preprocessing MRI data. The purpose of the work is to determine the effect of type 1 diabetes on the properties of the laboratory mice vasculature.

1. Introduction

Magnetic resonance imaging (MRI) is a non-destructive technique used to examine the inner structure of objects. The MRI is based on the Nuclear Magnetic Resonance (NMR) phenomenon, which is the rearrangement of magnetic moments of nonzero-spin nuclei in the external magnetic field. Employment of MRI in the investigation of bio-tissues and bio-fluids is based on NMR of hydrogen atoms. A specialized MRI technique called Magnetic Resonance Angiography (MRA) was developed to examine the circulatory system in living organisms [1]. Reconstructing the vasculature (and, in particular, that of the brain) of small laboratory animals in type 1 diabetes is of interest by virtue of applications of these results to deeper and more precise study of this severe disease. Animals genetic line and causing diabetes manipulation is briefly described in Section 2. MRA methods and reconstructed cerebral vasculature [2-4] is described in Section 3. Results of numerical analysis of the blood flow in the constructed models are obtained using software ANSYS/CFX and presented in Section 4. Results of statistical analysis of geometric configuration of blood net for the vessels of the Willis' circle and results of numerical analysis are carried out in Section 5.

2. Description of used animals

In this work used MRA angiograms obtained from male and female mice line NOD.CB17-Prkdcscid/J (NOD SCID). These mice were investigated in 14 weeks old. The diabetes group was injected intraperitoneally with streptozotocin (STZ; Sigma, USA) dissolved in 0.01 M citrate buffer at pH 4.2. A total of 150 mg STZ/kg body weight mass. The control group was injected with the same volume of the citrate buffer.

3. MRA methods and reconstructed cerebral vasculature

Non-fragmented three-dimensional models of vessel nets of all animals under investigation have been constructed using cylindrical RF coil MRI data. We use the software ITK-Snap [5] for segmentation. In Figure 1 you can see images of three-dimensional models of cerebral vessel nets for female animals of NOSCID genetic line, in Figures 2 — for male animals. Hereafter we denote Mxx male animals and Fxx female ones. The suffix "d" is used to indicate a sick animals.

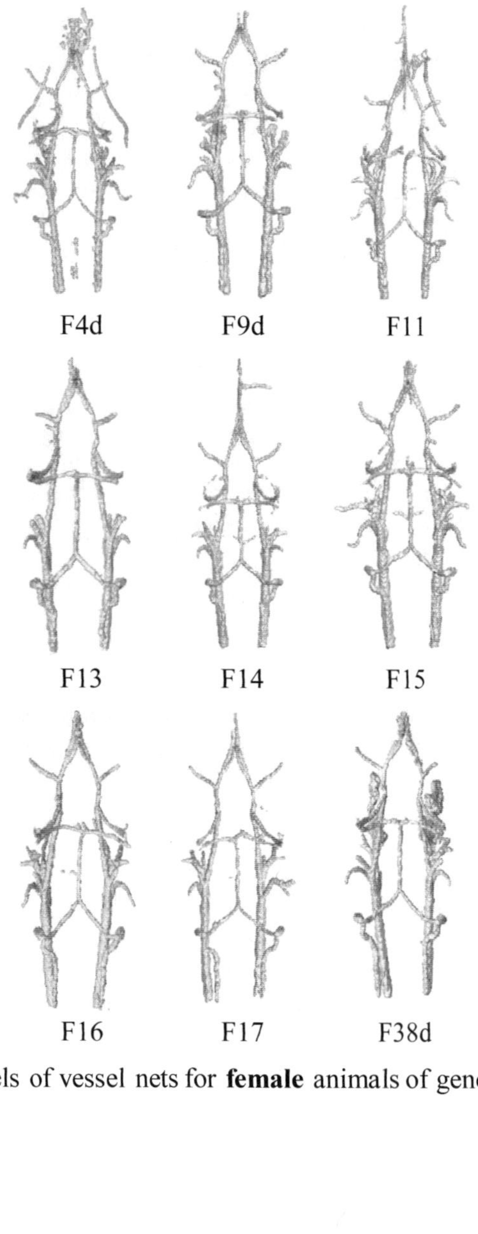

Figure 1. Models of vessel nets for **female** animals of genetic line NOSCID.

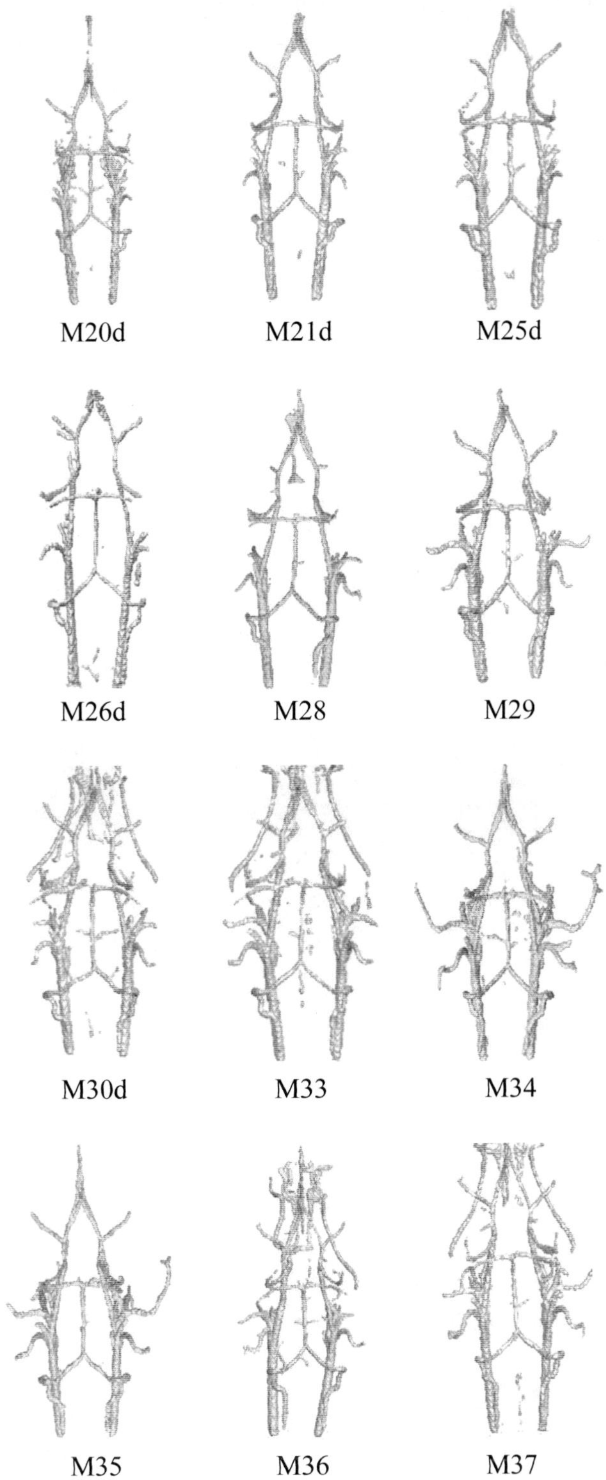

Figure 2. Models of vessel nets for **male** animals of genetic line NOSCID.

The next stage is the trimming and smoothing of the vasculature. Examples of the resulting vascular network in the vicinity of the circle of Willis are shown in Figure 3

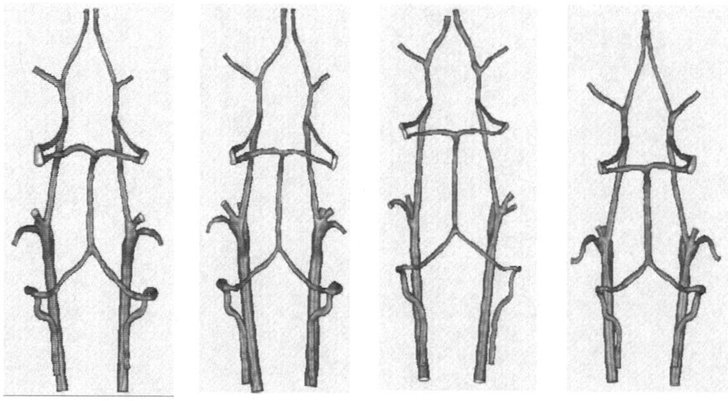

Figure 3. Examples of trimmed and smoothed vasculature.

At the received configurations of all mice vascular networks were measured geometrical parameters. Namely, the angles between the vessels of Willis' circle and sections of vessels in selected positions. Measured angles of the circle of Willis are shown in Figure 4.

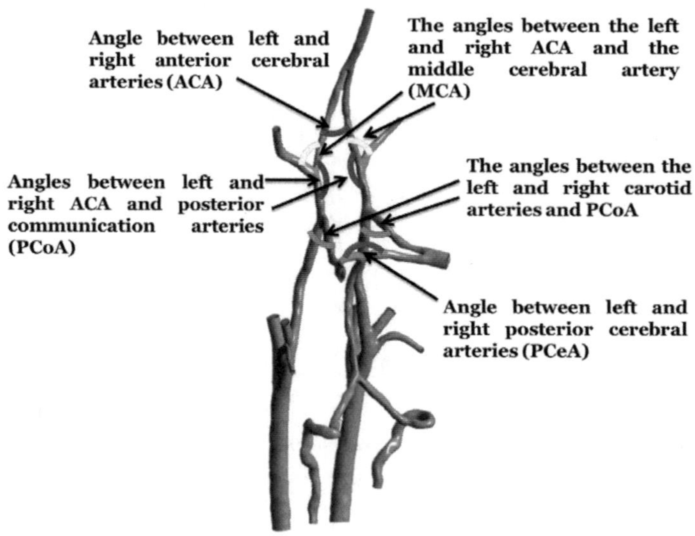

Figure 4. Measured angles of the circle of Willis.

Results of measurements of angles between the vessels of Willis' circle (being the part of build three-dimensional models) shown in Table 1. In the table rows correspond to condition numbers of investigated animals. The values of angles between blood vessels in places of their confluence adduce in columns. The angles between left and right anterior cerebral arteries (ACA) are adduced in the second column. The angles between left and right ACA and posterior communication arteries (PCoA) are adduced in the third and fourth columns. The angle between left and right posterior cerebral arteries (PCeA) is adduced in the fifth column. The angle between left and right anterior cerebral arteries (ACA) and middle cerebral artery (MCA) is adduced in the sixth and seventh column. The angle between left and right carotid arteries (CA) and posterior communication arteries (PCoA) is adduced in the eighth and ninth column.

Selected sections of vessels are shown in Figure 5.

Table 1. Angles of Willis' circle (in degrees).

Animal	ACA	ACA and PCoA (l.)	ACA and PCoA (r.)	PCeA	ACA and MCA (l.)	ACA and MCA (r.)	CA and PCoA (l.)	CA and PCoA (r.)
F4d	41	157	162	126	80	86	53	53
F9d	48	162	161	131	86	78	55	42
F11	46	165	164	165	115	64	57	59
F13	46	160	170	157	87	82	43	66
F14	23	163	165	134	68	79	58	55
F15	39	164	164	146	44	74	61	58
F16	39	162	160	145	88	76	53	57
F17	35	159	161	124	65	64	51	58
F38d	55	161	158	139	79	57	49	58
M20d	46	166	158	135	72	57	57	55
M21d	50	163	151	125	59	72	54	50
M25d	35	166	165	118	54	56	62	51
M26d	64	151	156	140	86	57	58	54
M28	36	160	152	133	73	114	62	55
M29	38	168	160	94	56	61	51	45
M30d	37	160	161	142	65	78	54	60
M33	42	167	157	114	76	75	57	51
M34	37	162	151	128	58	76	52	53
M35	42	163	160	129	66	64	58	57
M36	50	160	158	132	67	52	47	54
M37	41	163	151	137	69	57	55	57

4. Numerical analysis of hemodynamics

Software ANSYS/CFX is used for numerical analysis of hemodynamics [6, 7]. Computation carried out on the base of supercomputer centre of Novosibirsk State University. Model of incompressible viscous liquid (viscosity and density of liquid is given equal to viscosity and density of blood) is used for numerical modelling. Walls of blood vessels considered to be rigid and no-slip condition is given to them. Flow considered to be laminar, so the mathematical model of the blood flow consists of four equations expressing conservation laws of mass and three pulse component.

At the inlets of vertebral arteries and common carotid arteries we set blood flowrate, obtained on the basis of NMR phase velocimetry. In outlets we set volumetric rate different from the vessel outlet in the cross-section area.

As a result of numerical calculations were obtained values of velocity and pressure in the

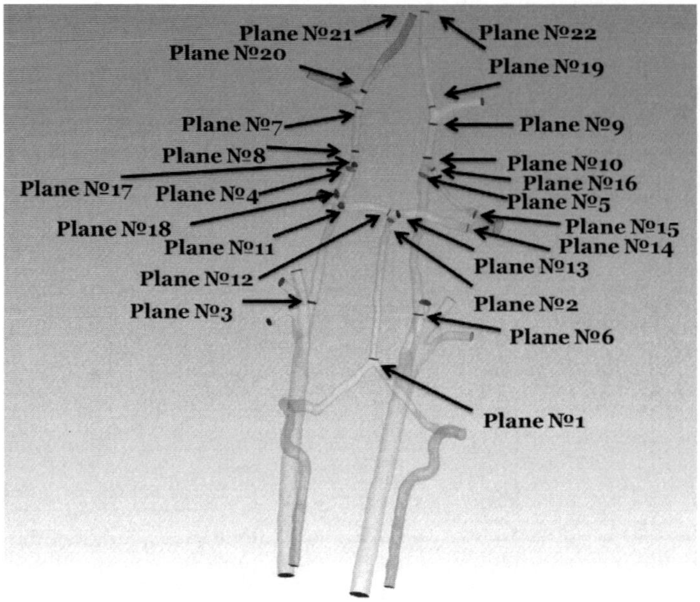

Figure 5. Selected sections of vasculature.

vasculature. Examples of these results of numerical simulations are shown below. In Figures 6, 7 you can see stream lines and variations in mean pressure distribution of the vessel wall near Willis' circle respectively.

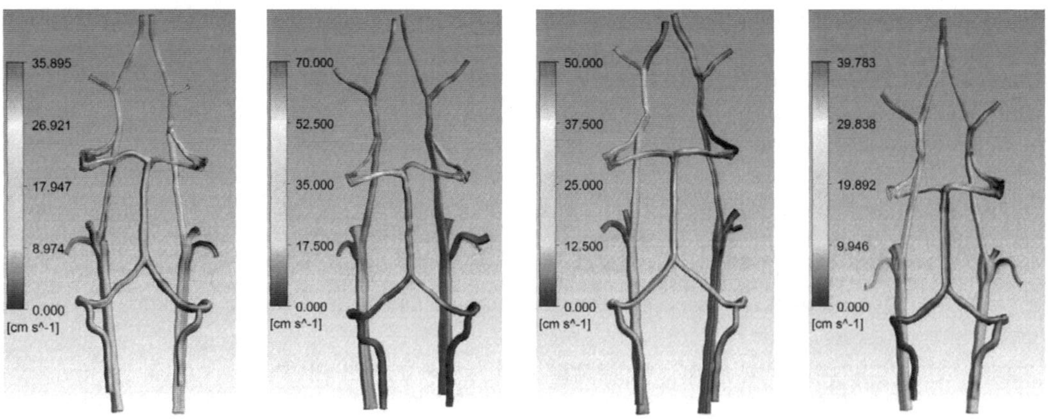

F4d F17 M26d M37

Figure 6. Stream lines of the velocity vector.

5. Statistical analysis

Using results of numerical calculations we found the values of flowrate, average pressure and maximum speed for all inputs, outputs, and highlighted in Figure 5 cross sections.

Also, the hydraulic resistances of all parts of the bloodstream were computed between selected sections in Figure 5.

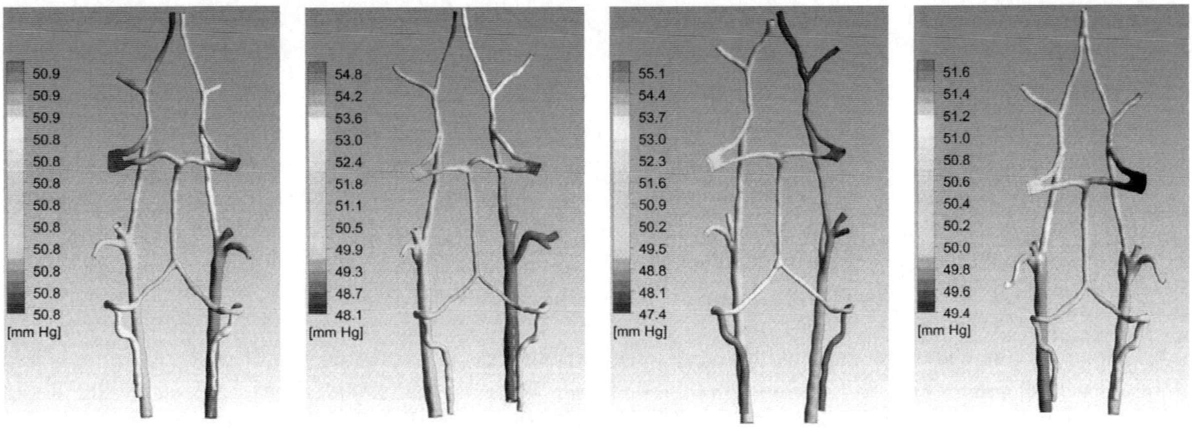

F4d F17 M26d M37

Figure 7. Variations from mean pressure distribution on the vessel wall.

All data in the form of averages by groups are presented in the following Table:

	female			male		
	control mean ± S.E.	sick mean ± S.E.	p	control mean ± S.E.	sick mean ± S.E.	p
$R^{M,1-2}$	187.19 ± 48.76	174.56 ± 40.86	0.87	130.23 ± 27.48	119.43 ± 41.01	0.82
$R^{L,3-4}$	49.43 ± 5.21	66.24 ± 4.08	0.08	48.25 ± 7.27	61.62 ± 11.59	0.33
$R^{R,5-6}$	55.75 ± 6.35	81.99 ± 9.13	0.05	67.28 ± 11.04	86.36 ± 32.02	0.53
$R^{L,7-8}$	25.09 ± 6.67	37.92 ±2.63	0.24	27.52 ± 7.40	17.56 ± 4.74	0.33
$R^{R,9-10}$	49.05 ± 12.68	40.83 ± 4.10	0.67	30.21 ± 4.90	43.67 ± 21.13	0.48
$R^{L,11-12}$	164.40 ± 45.02	81.52 ± 17.93	0.26	103.91 ± 25.41	166.62 ± 80.22	0.41
$R^{R,13-14}$	84.16 ± 13.99	171.44 ± 71.48	0.13	98.12 ± 34.67	95.08 ± 62.08	0.96
$R^{R,15-16}$	139.05 ± 28.50	235.86 ± 40.59	0.09	52.92 ± 44.91	139.36 ± 99.64	0.40
$R^{L,17-18}$	128.69 ± 38.46	158.49 ± 55.59	0.67	98.84 ± 51.15	23.52 ± 103.57	0.49
$R^{R,19-22}$	201.16 ± 35.10	158.90 ± 37.52	0.48	129.71 ± 21.39	147.42 ± 20.35	0.58
$R^{L,20-21}$	268.32 ± 110.74	212.14 ± 57.29	0.75	118.98 ± 21.82	124.15 ± 27.83	0.88
ΣR^L	635.94 ± 167.00	556.30 ± 56.31	0.76	397.50 ± 62.69	393.46 ± 189.48	0.98
ΣR^R	529.17 ± 68.84	689.01 ± 92.63	0.22	378.24 ± 37.45	511.89 ± 153.88	0.35
$\Sigma R^L - \Sigma R^R$	106.77 ± 107.60	-132.71 ± 75.38	0.19	19.27 ± 49.83	-118.42 ± 81.85	0.16
ΣS^L	0.65 ± 0.06	0.53 ± 0.03	0.23	0.68 ± 0.05	0.63 ± 0.08	0.54
ΣS^R	0.56 ± 0.04	0.50 ± 0.02	0.32	0.61 ± 0.03	0.59 ± 0.07	0.79
$\Sigma S^L - \Sigma S^R$	0.08 ± 0.04	0.03 ± 0.05	0.47	0.07 ± 0.03	0.03 ± 0.04	0.44
ΣQ^L	0.02 ± 0.01	0.03 ± 0.00	0.61	0.04 ± 0.01	0.03 ± 0.01	0.36
ΣQ^R	0.02 ± 0.01	0.01 ± 0.00	0.69	0.01 ± 0.01	0.02 ± 0.01	0.60
$\Sigma Q^L - \Sigma Q^R$	0.00 ± 0.02	0.02 ± 0.00	0.64	0.03 ± 0.02	0.01 ± 0.00	0.38
ΣV^L_{max}	216.06 ± 49.46	119.09 ± 11.76	0.22	187.70 ± 22.42	162.49 ± 25.18	0.48
ΣV^R_{max}	208.34 ± 55.69	135.05 ± 10.35	0.40	148.99 ± 12.93	177.99 ± 35.62	0.40
$\Sigma V^L_{max} - \Sigma V^R_{max}$	7.72 ± 22.91	-15.96 ± 8.91	0.51	38.70 ± 32.35	-15.50 ± 11.52	0.20

In this Table we denote R – vascular hydraulic resistance, $R[10^3 \cdot Pa \cdot s/m^3] = \Delta P/Q$, where ΔP – pressure difference; Q – volume flow rate; S – vascular cross-section area, $[mm^2]$; Q – blood volume flow rate $[cm^3/s]$; V_{max} – maximum blood velocity, $[cm/s]$.

Upper indices denote: R – right vasculature, L – left vasculature, M – medium vasculature. Number indices denote numbers of cross-section planes highlighted in Figure 5. The Σ sign denote sum of certain parameters on all left or right vasculature.

Statistical analysis has performed on the basis of the Student's t-criterion.

Using these data and the geometric characteristics of the vasculature. Statistical analysis was performed which showed that mainly significant differences between the control group of animals and sick one not identified. But in group of sick animals there is a otherwise relationship of indicators in the right and left vasculature in comparison with the control group. This applies primarily to summary resistance and summary maximum speed.

6. Conclusions

In the paper the construction of the models of vascular channel for the laboratory animals with and without diabetes are studied. MRI data is it possible to construct models of the vascular system, suitable for hemodynamic calculations and statistical analysis of vascular architectonics.

It was shown that there is influence of diabetes on morphological and hydrodynamical characteristics of Willis' circle of adult animals is weakly expressed and manifests itself only in the another relationship of indicators in the right and left vasculature.

Acknowledgments

The study was completed thanks to the Centre for Laboratory Animal Genetic Resources, Institute of Cytology and Genetics, Siberian Branch of the Russian Academy of Sciences (RFMEFI61914X0005 and RFMEFI62114X0010 – all MRI experimentation studies using Bruker BioSpec 117/16 USR scanner); the financing for the state task 0324-2016-0002 (acquisition of laboratory animals and the simulation of diabetes); the support of grant of the government of the Russian Federation (project 14.W03.31.0002 – mathematical modeling).

References

[1] Beckmann N 2000 *Magn. Reson. Med.* **44** 252–8

[2] Cherevko A A, Yankova G S, Maltseva S V, Parshin D V, Akulov A E, Khe A K and Chupakhin A P 2016 *Journal of Physics: Conference Series* **722** 012029 1–7

[3] Maltseva S V, Cherevko A A, Khe A K, Akulov A E, Savelov A A, Tulupov A A, Derevtsov E Yu, Moshkin M P and Chupakhin A P 2016 *Appl. Magn. Reson.* **47** 23–39

[4] Angenent S, Pichon E and Tennenbaum A 2006 *B. Am. Math. Soc.* **43** 365–96

[5] Yushkevich P A, Piven J, Hazlett H C, Smith R G, Ho S, Gee J C and Gerig G 2006 *Neuroimage* **31** 1116–28

[6] Vorobtsova N A, Yanchenko A A, Cherevko A A, Chupakhin A P, Krivoshapkin A L, Orlov K Yu, Panarin V A and Baranov V I 2013 *Russ. J. Numer. Anal. M.* **28** 505–16

[7] Yanchenko A A, Cherevko A A, Chupakhin A P, Krivoshapkin A L and Orlov K Yu 2014 *Russ. J. Numer. Anal. M.* **29** 307–17